BONES

AND

BODIES

BONES

—— AND ——

BODIES

How South African Scientists Studied Race

ALAN G. MORRIS

WITS UNIVERSITY PRESS

Published in South Africa by:
Wits University Press
1 Jan Smuts Avenue
Johannesburg 2001

www.witspress.co.za

First published 2022

http://dx.doi.org.10.18772/12022027236

978-1-77614-723-6 (Paperback)
978-1-77614-724-3 (Hardback)
978-1-77614-725-0 (Web PDF)
978-1-77614-726-7 (EPUB)

Project manager: Alison Lockhart
Copyeditor: Alison Lockhart
Proofreader: Lee Smith
Indexer: Sanet le Roux
Cover design: Hybrid Creative
Typeset in 11.5 point Minion Pro

CONTENTS

A NOTE ON THE USE OF HISTORICAL TERMINOLOGY   vii

LIST OF ILLUSTRATIONS   ix

ACKNOWLEDGEMENTS   xi

LIST OF CHARACTERS WITH DATES OF BIRTH,
DEATH AND AFFILIATION   xvi

SCHEMA OF TYPES   xix

INTRODUCTION   1

CHAPTER 1   Dr Louis Péringuey's Well-Travelled Skeletons   11

CHAPTER 2   Boskop: The First South African Fossil Human Celebrity   39

CHAPTER 3   Matthew Drennan and the Scottish
Influence in Cape Town   65

CHAPTER 4   The Age of Racial Typology in South Africa   108

CHAPTER 5   Raymond Dart's Complicated Legacy   144

CHAPTER 6   Ronald Singer, Phillip Tobias and the
'New Physical Anthropology'   191

CHAPTER 7   Physical Anthropology and the Administration
of Apartheid   247

CHAPTER 8   The Politics of Racial Classification in
Modern South Africa   275

SELECT BIBLIOGRAPHY   297

INDEX   325

A NOTE ON THE USE OF HISTORICAL TERMINOLOGY

The terminology describing South African peoples has changed dramatically over the years. Modern usage has tried, where possible, to choose language that is acceptable to the descendant communities themselves, as many of the earlier terms are now considered insulting at best and virulently racist at worst.

Which modern terminology is most appropriate? This is indeed a difficult question as some names are acceptable in one location or context, but not in another. I have tried to use as a guide the nomenclature of southern African people as proposed at the June 1971 Royal Society of South Africa meeting (Jenkins and Tobias 1977). This meeting suggested that separate terminology be used for discussions of biology, language and culture/economy. Wherever possible, I have tried to use names based on languages spoken by modern or historical populations. Linguistic terms have a major advantage in that they are often ethnically based and self-defined, but they do not always overlap with biological origin. People speaking languages from entirely different families can have strong biological relationships and vice versa. Although the old racial categories of 'Caucasoid', 'Mongoloid' and 'Negroid' are considered to have limited value in categorising biology, I have still used the terms 'South African Negro' and 'Khoesan', as they gather people into two biological clusters; the former being linguistically related Bantu-speaking populations, while the latter is made up of people speaking languages from at least three completely different families, yet sharing significant biological similarity.

I have not replaced old terms with their modern equivalents where the results of earlier research are presented, as the modern and older terms are not always interchangeable. For example, Khoekhoen today is accepted as an ethnic term referring to people who are culturally defined as speaking a Khoe language (Jones 2003), but a century ago the term 'Hottentot' had both biological and political meanings, often referring to people of mixed genetic and cultural heritage. It is thus important that when reporting on the early research we use the terms as those researchers did, only bringing the terminology into modern form when we are summarising the early research in relation to what we know today.

Therefore, much of the early terminology is retained in this book as it was used in the research that is being discussed. Where terms are no longer acceptable, the historical use is marked by the employment of quotation marks, not to emphasise the validity of the terms, but as a convention to show that the terms today are problematic. This includes both ethnic and racial terms.

LIST OF ILLUSTRATIONS

Figure 1.1   Staff of the South African Museum in 1920   19

Figure 2.1   The cranium of the Boskop specimen   41

Figure 2.2   A photograph of Frederick William FitzSimons   44

Figure 2.3   Frederick FitzSimons excavating at Whitcher's Cave,
              near George, *circa* 1921   49

Figure 2.4   An unnamed cave in the Tsitsikamma region
              excavated by Frederick FitzSimons in the 1920s   62

Figure 3.1   Matthew Drennan in his anthropology laboratory at
              the University of Cape Town in 1931   73

Figure 3.2   The Cape Flats skull, discovered by Matthew
              Drennan near Cape Town in 1929   87

Figure 3.3   Donkey cart access to the Elandsfontein fossil
              site near Hopefield, *circa* 1952   95

Figure 3.4   Tea break during excavations at Elandsfontein
              in the early 1950s   96

Figure 3.5   The Saldanha skull after reconstruction   97

Figure 3.6   E.N. (Ted) Keen holding stone and bone artefacts
              at the Elandsfontein fossil site, *circa* 1953   98

Figure 3.7   Matthew Drennan at the Elandsfontein
              fossil site, *circa* 1953   100

Figure 4.1   Thomas Frederick Dreyer, 1937   113

Figure 4.2   Thomas Dreyer's reconstruction of the Florisbad hominin   118

Figure 4.3   Vernon Brink with other unnamed medical students
at Oxford University, *circa* 1921   122

Figure 4.4   Illustration from Thomas Dreyer and Albert Meiring's
1937 article on the Kakamas burials   128

Figure 5.1   Raymond Dart in 1925 with the Taung skull shortly
after its discovery   151

Figure 5.2   Lawrence Wells, Alexander Galloway and Trevor
Trevor-Jones studying the Bambandyanalo
skeletons in 1936   169

Figure 6.1   Robert Broom at the Sterkfontein fossil site in
August 1936   198

Figure 6.2   The members of the 1951 Panhard-Capricorn
Expedition just before their departure for the Kalahari   201

Figure 6.3   Excavating Griqua graves in April 1961 at the
old Campbell cemetery in the Northern Cape   238

Figure 6.4   Raymond Dart's 84th birthday celebration at the
Department of Anatomy, University of the
Witwatersrand, February 1977   240

ACKNOWLEDGEMENTS

Before anyone else, I need to start by thanking my wife, Liz, for putting up with my bad temper and the single-minded focus that always accompanies my writing. Honey, I am sorry that I did it to you again! Once more, you came to the fore in reading through my scramble of pages and giving me the encouragement to continue – oh, and sometimes warning me that it was not working and that I needed to go back to the keyboard. With this book, I occasionally roped in my daughter Leigh as well. I really appreciate both of you.

Other than my stalwart wife and daughter, there has been a whole range of people who have given me their time to answer questions, provide guidance or give me access to unpublished material. Since each chapter in this book is a snapshot of particular people, the best way to acknowledge all of this help is chapter by chapter.

1. Dr Louis Péringuey's well-travelled skeletons

I would like to thank Wendy Black, Wilhelmina Secunda, Liesl Ward and Lailah Hisham for permission to work on the Eugène Pittard files at the Iziko Museum and especially to Baheya Hardy and Lulama Noduma in the Social History Library. Marie Joseph O'Connor and Violette Kramer went way beyond the call of duty to help with translations from the original French. Tracking down information about Frank Shrubsall was especially difficult and Alexandra Browne, the archivist at Clare College, Cambridge, was most helpful.

2. Boskop: The first South African fossil human celebrity

The Boskop story involved tracking Louis Péringuey and Frederick FitzSimons in Cape Town and Port Elizabeth, respectively. Wendy Black, Wilhelmina Secunda, Liesl Ward and Lailah Hisham again helped with the correspondence files at the Iziko Museum. Nancy Teitz, Mike Raath, Dorothy Pittman and Emile Badenhorst all were exceptionally helpful in tracking down museum minutes, short biographies, newspaper articles and old pictures of the distinctive Mr FitzSimons. I am indebted to Chris Stringer at the Natural History Museum in London for warnings about the current Boskop publicity and to Fred Grine for his comments on an earlier draft of the chapter.

3. Matthew Drennan and the Scottish influence in Cape Town

Writing anything about Matthew Drennan would not have been possible without the help of Caroline Powrie in the Department of Human Biology at the University of Cape Town (UCT). Caroline literally knew where the bodies were buried and helped me to find all the old boxes with the detritus of Drennan's long sojourn in the department. She not only found the wonderful old pictures of Drennan, but also his boxes of newspaper clippings, manuscripts, letters and old articles. Michael Cassar and Charles Slater of UCT and Brendon Billings from the University of the Witwatersrand (Wits) helped with information about cadavers. I very much appreciate Howard Phillips from the Department of History at UCT, who alerted me to John Goodwin's relations with Keith Jolly and Drennan. Andrea Walker of UCT Libraries helped retrieve Goodwin's letters. Vida Milovanovic of the Royal College of Surgeons archives in London and Malcolm MacCallum and Ruth Pollitt of the Anatomical Museum in Edinburgh helped with some of Drennan's and Lawrence Wells's letters in their collections. Vicky Gibbons of the Department of Human Biology at UCT kindly read an earlier draft of this chapter and helped with comments and direction.

4. The age of racial typology in South Africa

Liz de Villiers, Sudré Havenga, Zoë Henderson and the late James Brink, all of the National Museum in Bloemfontein were exceptionally helpful in finding out about Thomas Dreyer's activities. Thanks especially to James for sharing Dreyer's unpublished manuscripts. Tracking the elusive Vernon Brink was particularly challenging and I am most grateful to Rob Kruzinski in London, David Morris at the McGregor Museum, Kimberley, Handri Walters in Stellenbosch and Howard Phillips from the Department of History at UCT for providing clues to his origin and activities. Handri also gave me access to her newly minted PhD thesis, which was tremendously helpful. Goran Štrkalj, of the University of New South Wales, Australia, shared with me Robert Broom's letter that he found in the American Museum in New York. My thanks also go to Saul Dubow for guidance on the events of the first half of the twentieth century.

5. Raymond Dart's complicated legacy

Elizabeth Marima, the archivist at Wits Library, kindly provided access to the Raymond Dart and Phillip Tobias files. Goran Štrkalj forwarded a copy of Dart's travelogue from his files, and Raoul Coscia of Cape Town helped immensely with translations from Italian in Nino del Grande's letters. Jonathan Reinarz of the University of Birmingham helped to identify Robert Lockhart as the successful candidate in the Birmingham job in 1931, and Heather Kennedy of the University of Aberdeen provided a list of Lockhart's publications, so that I could compare his qualifications with Dart's.

6. Ronald Singer, Phillip Tobias and the 'new physical anthropology'

Jason Hemingway provided information about Tobias's students and his supervision. Bev Kramer was kind enough to confirm some of my impressions of the difficulties in dealing with Tobias on the employment horizon. Laurel Baldwin-Ragaven and Leslie London helped to track down an elusive reference from UCT's Faculty of Health Sciences. Ron Singer spoke little about his childhood, so it was a great relief when I was able to contact his daughter, Hazel Singer Griffiths, through the kind offices of Michael Solomons in Cape Town. Charles Musiba at the University of Colorado at Denver, Curtis Marean at Arizona State University in Phoenix and Callum Ross at the University of Chicago all helped in tracing Singer's papers after his death in 2006.

7. Physical anthropology and the administration of apartheid

Most of the information about the development and management of apartheid is available in the literature, but Yvonne Erasmus was very helpful in providing me with access to her unpublished PhD thesis, which discussed and digested this material. Attie Tredoux, the chief legal adviser of Parliament for many years, not only allowed me to interview him, but also gave me his own copies of Home Affairs reports from the 1980s. Aron Mazel, currently of Newcastle in the United Kingdom, provided some of the Cultural History Museum correspondence that helped to explain Drennan's personal views on the politics of his time as a museum board member.

* * *

Wits University Press sent the manuscript to three anonymous reviewers, who offered excellent comments and provided really useful ideas for restructuring the text. Of course, the team at Wits University

Press – Veronica Klipp, Roshan Cader and Kirsten Perkins – deserves credit for guiding the whole process of getting this from manuscript to book. My thanks also to project manager and editor, Alison Lockhart; Karen Lilje of Hybrid Creative for her work on the images and cover; Lee Smith for proofreading and Sanet le Roux for the index.

Last, but definitely not least, I want to thank all the people I interviewed over the years. Seven of them – Ralph Ger, Margaret Shaw, Attie Tredoux, Alun Hughes, Ted Keen, Hertha de Villiers and Ron Singer – sat with me in front of my tape recorder and spoke openly about their own experiences. Others, including Geoff Sperber, George Nurse, Marie Barry, Phillip Tobias and Ralph Kirsch, spoke to me in a less formal environment, but all of them were fantastic in their willingness to talk about things past. Without them, this book could not have been written.

LIST OF CHARACTERS WITH
DATES OF BIRTH, DEATH AND AFFILIATION

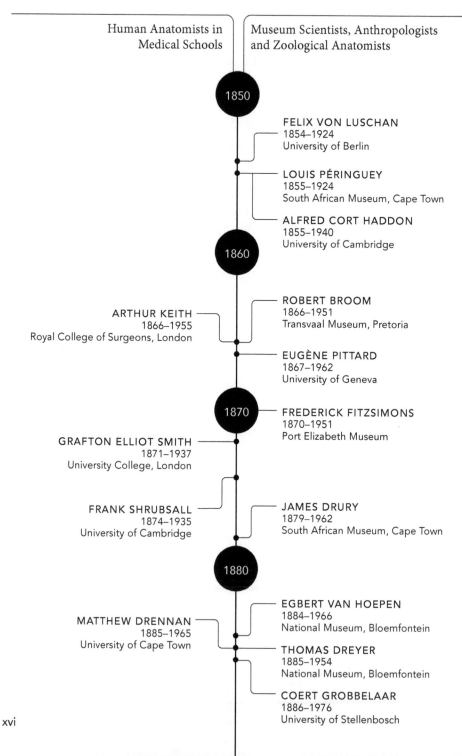

Human Anatomists in
Medical Schools

Museum Scientists, Anthropologists
and Zoological Anatomists

1850

FELIX VON LUSCHAN
1854–1924
University of Berlin

LOUIS PÉRINGUEY
1855–1924
South African Museum, Cape Town

ALFRED CORT HADDON
1855–1940
University of Cambridge

1860

ROBERT BROOM
1866–1951
Transvaal Museum, Pretoria

ARTHUR KEITH
1866–1955
Royal College of Surgeons, London

EUGÈNE PITTARD
1867–1962
University of Geneva

1870

FREDERICK FITZSIMONS
1870–1951
Port Elizabeth Museum

GRAFTON ELLIOT SMITH
1871–1937
University College, London

FRANK SHRUBSALL
1874–1935
University of Cambridge

JAMES DRURY
1879–1962
South African Museum, Cape Town

1880

EGBERT VAN HOEPEN
1884–1966
National Museum, Bloemfontein

MATTHEW DRENNAN
1885–1965
University of Cape Town

THOMAS DREYER
1885–1954
National Museum, Bloemfontein

COERT GROBBELAAR
1886–1976
University of Stellenbosch

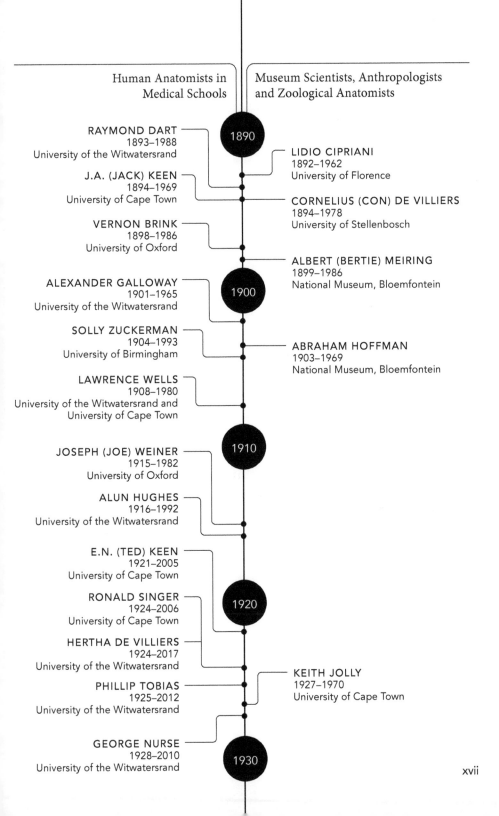

Human Anatomists in Medical Schools

Museum Scientists, Anthropologists and Zoological Anatomists

RAYMOND DART
1893–1988
University of the Witwatersrand

LIDIO CIPRIANI
1892–1962
University of Florence

J.A. (JACK) KEEN
1894–1969
University of Cape Town

CORNELIUS (CON) DE VILLIERS
1894–1978
University of Stellenbosch

VERNON BRINK
1898–1986
University of Oxford

ALBERT (BERTIE) MEIRING
1899–1986
National Museum, Bloemfontein

ALEXANDER GALLOWAY
1901–1965
University of the Witwatersrand

SOLLY ZUCKERMAN
1904–1993
University of Birmingham

ABRAHAM HOFFMAN
1903–1969
National Museum, Bloemfontein

LAWRENCE WELLS
1908–1980
University of the Witwatersrand and
University of Cape Town

JOSEPH (JOE) WEINER
1915–1982
University of Oxford

ALUN HUGHES
1916–1992
University of the Witwatersrand

E.N. (TED) KEEN
1921–2005
University of Cape Town

RONALD SINGER
1924–2006
University of Cape Town

HERTHA DE VILLIERS
1924–2017
University of the Witwatersrand

KEITH JOLLY
1927–1970
University of Cape Town

PHILLIP TOBIAS
1925–2012
University of the Witwatersrand

GEORGE NURSE
1928–2010
University of the Witwatersrand

SCHEMA OF TYPES

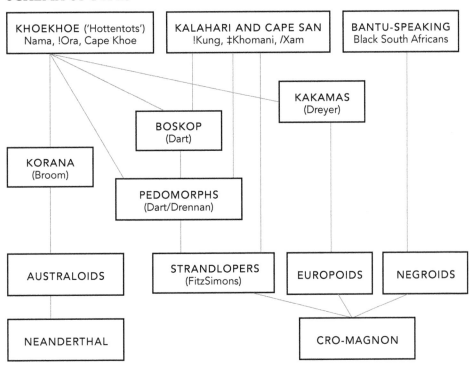

INTRODUCTION

Why is it important that we as scientists study our own history? The aphorism first spoken by the American philosopher George Santayana (and paraphrased by Winston Churchill) is especially true for physical anthropology: 'Those who cannot remember the past are condemned to repeat it.' The subject's past is not a pleasant one. Physical anthropology, the branch of anthropology that considers the structure and evolution of the human body, has been used to justify slavery, condemn criminals by their appearance, limit immigration according to racial origin and, in the case of Nazi Germany, to commit genocide. Over the years, South African physical anthropologists have written a great deal about the peoples of southern Africa and we need to ask if these publications have contributed to our own social heresies. That, of course, will be the task of historians, but we need to be aware that the old problems continue to surface. The publication of the book *The Bell Curve* (Herrnstein and Murray 1994) and the writings of Phillippe Rushton (for example, 1995) have tried to resurrect biological racism by stratifying levels of intelligence by race. They are aberrations that have triggered heated responses from professional physical anthropologists, but in the eyes of the public such ideas do have legitimacy. In the South African context, despite having vanquished the apartheid dragon, we need to understand exactly how much of the racist underpinnings of the policy have become internalised and are still part of us. Academics struggle to find ways to balance the roles of sociology and genetics in their research and the lay public continues to perceive people in neat racial categories. Political attempts to stamp out the old racism against black people in South Africa are ongoing, but sometimes that is matched, either intentionally or unintentionally, by reverse racism. Understanding the past will guide us in how to solve these problems in the future.

Jonathan Marks wrote an extremely valuable book in 2009, *Why I Am Not a Scientist: Anthropology and Modern Knowledge*, which looks at the intersection between anthropology and science. Despite the title of the book, Marks is not anti-science and contends that he can be a positive critic because anthropology bridges both science and the humanities. He maintains that scientists fool themselves into failing to see how their ideas are influenced by events and ideas around them. Throughout my career I have balanced my research on human skeletal variation with the knowledge that the skeletons I study were once the frames of living people and that my biological information can have no value without understanding the social and cultural context of the people that I study. I want to look at the history of my subject to show how I have been sensitive to how history has impacted on my research and conclusions.

Anthropological discoveries in South Africa over the past century have been of exceptional importance in terms of our understanding of human evolution, but also because they have influenced our society in ways that have not always been positive. It has been the anatomists in the medical schools who have most influenced our understanding of human structure and variation, and their racial classifications and descriptions of the peoples of southern Africa have flowed into medical specialities, including surgery, gynaecology, forensics, genetics and epidemiology/public health. The same anatomists who have dabbled in physical anthropology have also taught racial variation to generations of medical students at undergraduate and postgraduate level. My choice of the word 'dabbled' is intentional, as none of these scholars were trained in the discipline of anthropology, yet generations of researchers in medical, natural and social sciences have used their classifications and categories. From this perspective, understanding the historical basis of physical anthropology becomes critically important.

In 1991, when I was just becoming interested in the history of science, I submitted an abstract for a conference paper on the history of physical anthropology to the American Association of Physical Anthropologists titled 'The Philosophical Roots of the Study of Human Variation in South Africa'. I was keen to understand the motivations of the earlier researchers, especially in relation to their acceptance of the typological model of race. To my surprise, the abstract was rejected because it 'lacked originality'. The reviewer said that most of this had been covered by Phillip Tobias's history of physical anthropology (Tobias 1985a). Indeed, I suspect the reviewer had not actually read Tobias's paper, as it entirely lacks any information about the researchers themselves. Tobias's paper was a masterful account of who did what, where and when, but it lacked any information about the historical characters and their motivations. So, in a sense, this book is the conference paper that never was.

My training and my career are overwhelmingly in physical anthropology, not history. My doctoral thesis examined a series of archaeologically derived human skeletons from the late seventeenth to the early nineteenth centuries excavated from locations along the historical border of the Northern Cape Colony (Morris 1984). The Northern Cape frontier was a zone of contact between Khoekhoe herders, Sesotho- and Setswana-speaking agriculturalists and people of mixed origins from the Cape Colony. The skeletons from the four sites were held in the McGregor Museum in Kimberley and the National Museum in Bloemfontein. The boxes containing the skeletons were brought to my office in Johannesburg, where the bones were meticulously cleaned, reconstructed and studied. In order to make sense of the skeletal variation seen in the archaeological skeletons, it was necessary to find modern skeletons from related populations for comparison. What became obvious was that the skeletons accessioned in many of the museum and medical school collections were identified not on the basis of known self-defined ethnicity, but were lodged there as racial types determined by the accumulators and managers of the collections. What I had hoped to find was neat assemblies of individuals donated through medical school dissection programmes and recovered from community cemeteries, but the reality was very different. Many of the skeletons of people who had been known in life were labelled according to a strict racial typology in which racial identity was based on appearance, not the culture or the community from which he or she originated. This opened the world of skeleton collecting to me and brought a context to the old bones in the boxes. I found I had to deconstruct what was said in the register and try to dig back into the history of the person and how they had come to be in a modern collection of human skeletons. Something that had started as a search for ethnically identified skeletons grew into a much larger project looking at the origins of the collections themselves. What became apparent was how involved the physical anthropologists were as collectors and how ingrained their method of typology had become in the collection and description of 'specimens' and in their publications. Eventually, I examined more than 30 different collections and looked at the records of more than 2 500 human skeletons. This research was published as *A Master Catalogue: Holocene Human Skeletons from South Africa* (Morris 1992a). In the end, I rejected all skeletons that had been identified by ethnicity or race and accepted only those that could be associated with an archaeological context or had firm radiocarbon dates that placed them securely in time.

My archaeological interests drew me to the newly formed Centre for African Studies at the University of Cape Town (UCT). At one of their seminars, I was horrified to hear Professor Neville Alexander state that there was 'no such thing as race' and therefore there was no point in studying the differences between people. The shock to me was that someone who was steeped in social sciences could simply reject an entire field of study. This spurred me to present my own paper, 'The Role of Physical Anthropology in the Study of African History' (Morris 1980), to the Centre for African Studies. In it, I outlined why studies of human skeletons are not only valid, but also could provide exceptionally useful information for understanding history. This initiated my collaboration with Martin Hall of the Department of Archaeology at UCT and together we examined the problems that typological assessments had created in understanding the Iron Age populations in southern Africa (Hall and Morris 1983). For most archaeologists, the old typological ideas were still predominant in the literature and this was an opportunity to introduce more modern ideas of populations to archaeologists and sociologists who had not been exposed to this before.

Racial types had confused and misdirected archaeology, but they were even more insidious when it came to apartheid. The scientists themselves seemed to be unaware that their lack of comment on the absurdity of apartheid was in itself a statement and I pointed this out in a presentation to the Royal Society of South Africa (Morris 1988b). The Royal Society talk looked at the published articles in the *South African Medical Journal* for the year 1987 and demonstrated that the authors of these papers had almost no idea what they were writing about when it came to race. Biology and ethnicity were confused, generalisations about race were made from local samples and clinical papers described case studies as if there were no variation at all. The worst was a study of Gaucher disease in which the frequency of gene carriers was calculated for the Cape coloured population as enumerated in the 1980 census (Swart et al. 1987). The authors utterly failed to understand the political nature of the 'coloured' category and that the census numbers reflected multiple populations for whom a single frequency value had no meaning.

My research has always centred on the peopling of southern Africa and because of this I have always been sensitive to the issue of race and its meaning. When I joined the Department of Anatomy at UCT in 1981, I took on the unofficial role of department historian, especially with respect to things anthropological. This included storing boxes of old correspondence, lantern slides and old articles, but sorting through these had to wait until my retirement approached in 2014. When I reached retirement, it not only gave me the opportunity to begin to put more than 30 years of my research together, but it also was an opportunity to try to organise the historical material that was stored in the boxes in my office and around the department. Over a period of four months, a group of students, funded by the UCT Humanitec Digital Archive project, helped me to scan and organise the collection that will be accessible through the Department of Human Biology. The organisation of the collection provided the opportunity for me to tackle a final historical task – that of writing a single volume that would encompass this wealth of unpublished material.

My files on the historical characters in physical anthropology and related subjects had become extensive over the years. I tried to collect as many of the published books and papers as I could, but there is also a growing collection of less substantial documents. I met many of the earlier researchers: Lawrence Wells, Ronald Singer, Ted Keen, Raymond Dart and of course Phillip Tobias. But I also crossed the paths of many of the researchers who died before I arrived in South Africa and I have learned so much about them that I feel they are also people I have known: Louis Péringuey, Matthew Drennan, Robert Broom and Thomas Dreyer. Back in the early 1990s, I purchased a small tape recorder and sat down to talk with Alun Hughes, Margaret Shaw, Ted Keen, Hertha de Villiers and Ronald Singer, all of whom were either major players in physical anthropology or had worked closely with the earlier scientists who are now gone. I also made extensive notes on conversations that I had with many others. I managed to find a large number of old letters written either by these people or to them in the archives of my own Department of Human Biology in Cape Town and at the School of Anatomical Sciences in Johannesburg, but also at the Iziko Museum in Cape Town, the National Museum in Bloemfontein, and the Albany Museum in Grahamstown. More recently, I have tracked some old letters in Cambridge and London. These stories are my take on these individuals. I have an insider's view that may in itself be contested by future historians, but this is my chance to get my perspective down on paper.

This book consists of eight anthropological vignettes, each examining specific researchers or topics that had a special impact on South African physical anthropology. The book has been a long time in the making and it has drawn heavily on the unpublished Anatomy Archive stored in the Department of Human Biology at UCT, as well as the archives at the University of the Witwatersrand (Wits) and the interviews and museums mentioned above.

The first chapter focuses on Dr Louis Péringuey, the director of the South African Museum in the first two decades of the twentieth century. Péringuey engaged with Frank Shrubsall and Eugène Pittard to produce the first professional examination of archaeological skeletons from the growing collection in Cape Town. Sources for this chapter include their previously unpublished correspondence, along with the archive of Pittard's papers held at the Iziko Museum. The second chapter looks at the Boskop discovery and how a poorly preserved archaeological cranium directed the interpretation of human evolution in southern Africa for more than 50 years. In the same manner as the chapter on Péringuey, the story of Boskop draws on the unpublished correspondence archives at the Iziko Museum. The third chapter focuses on Matthew Drennan and the Department of Anatomy at the

UCT Medical School. Surprisingly, for someone who was a professor of anatomy at UCT for 40 years, almost nothing has been written about him. This chapter draws heavily on the UCT Department of Human Biology Anatomy Archive and explores Drennan's research activities and his motivations. Most important is the information drawn from interviews with Ted Keen and Ronald Singer, who both worked under him in the 1940s and 1950s. Chapter 4 moves away from the great centres of Cape Town and Johannesburg to look at the lesser-known physical anthropological activities in Bloemfontein, Kimberley and Stellenbosch. The emphasis in this chapter is on the work of Thomas Frederick Dreyer, Robert Broom and Gert Grobbelaar and the collection of human skeletons in the Karoo and Northern Cape during the 'age of racial typology'.

Chapter 5 zones in on the world-renowned Johannesburg anatomist Raymond Dart, his colleague Alexander (Sandy) Galloway and his student Lawrence Wells. Recent scholarship examining Dart's role in the development of physical anthropology in South Africa has been very critical (Dubow 1996; Derricourt 2009; Kuljian 2016). This chapter extends their work by including unpublished correspondence from Dart from the Wits archives and from Wells from the UCT Department of Human Biology Anatomy Archive, as well as from personal interviews with Alun Hughes and Hertha de Villiers, people who knew them well. The sixth chapter surveys the change in physical anthropology that came about in the 1950s and 1960s under the leadership of Ronald Singer in Cape Town and Phillip Tobias in Johannesburg. The arrival of the 'new physical anthropology' on the shores of South Africa is intimately connected with these two researchers and it created a new dynamic in scientific approach exactly at the time when the policy of apartheid was being implemented in the country. There is no shortage of information on Tobias (almost all of it written by himself) but winnowing out the key elements of his approach to research is more difficult. Most useful have been the interviews with Singer, Hughes and De Villiers, along with my own personal correspondence and letters, archived at Wits.

In the last two chapters, I do not deal with specific individuals, but instead look at the implementation of apartheid, how it affected the researchers, and how it continues to impact on physical anthropology in the present. Chapter 7 is constructed around a comment made by Tobias, claiming that 'no South African physical anthropologist was involved in providing the scientific underpinning for the government's policy of apartheid' (Tobias 1985a, 32). Since understanding apartheid is a prerequisite for evaluating the truth of Tobias's claim, much of this chapter looks at the classification system of apartheid and how it functioned. Included are previously unpublished descriptions of how scientists engaged with the classification system – for the most part as antagonists, but not entirely. Chapter 8 ends the volume by looking at issues of race classification in South Africa today and how it still plagues our lives as citizens and scientists.

1

Dr Louis Péringuey's Well-Travelled Skeletons

Louis Péringuey was an unlikely archaeologist. The director of the South African Museum in Cape Town from 1906 to his death in 1924, Péringuey identified himself as an entomologist, but in fact he had no formal education in science. Born of Basque parentage in Bordeaux in 1855, he went to war against the Prussians in 1870 without completing high school. Roger Summers suggests that his distress at the defeat of France in the war affected him greatly and he decided to 'seek his fortune outside of France' (Summers 1975, 56). What followed was several years of travelling through Senegal, The Gambia and Madagascar, with Péringuey finally arriving in South Africa in 1879. He obtained employment as a teacher of French, but it was at this time that he seems to have developed his lifelong interest in beetles, as a result of which he joined the South African Museum in 1884. His job as scientific assistant in the museum's Department of Entomology brought him into contact with the Phylloxera infestation that was decimating South Africa's vineyards in the 1880s. He became an inspector of vineyards in 1885 and was appointed colonial viticulturist in 1889, about which Director Trimen complained that the colonial government was using museum staff instead of setting up their own entomological service (Summers 1975, 57). Although these posts took him away from his entomological work at the museum, it was these new tasks that first exposed him to archaeology.

Péringuey excelled at his entomological work despite his lack of formal education. He published multiple papers on South African Coleoptera (beetles) between 1884 and 1906 and in 1907 he was awarded an honorary Doctor of Science (DSc) degree from the University of the Cape of Good Hope in appreciation of his scientific achievements (Summers 1975, 97; Plug 2020d).

Péringuey's work in the soil of vineyards revealed ancient stone tools and triggered an interest that would consume him for the rest of his career. In 1899 he, along with George Steuart Corstorphine of the South African College, made a major archaeological discovery of Palaeolithic artefacts similar in form to the Acheulian in Europe, which they termed 'Stellenbosch' after the district of their discovery (Péringuey and Corstorphine 1900, xxiv).

Péringuey's archaeology was breaking new ground in Africa. The concept of a series of technological ages – Stone, Copper, Bronze, Iron – had been recognised in Europe from the 1850s and the term 'Palaeolithic', or 'Old Stone Age', had been coined in the 1860s. This had been subdivided into Lower, Middle and Upper Palaeolithic based on stone tool technology. The rough hand-axes and cleavers made from large stone flakes and cobbles and shaped into bifacial (chipped on both sides) tools had been described from the Lower Palaeolithic site of Saint-Acheul in 1855, and it was this site that provided the name Acheulian. Further exploration of European arch-aeological sites in the latter half of the nineteenth century produced evidence of a much more complex stone tool technology, which was called Upper Palaeolithic. Whereas no human remains could be initially associated with Acheulian sites, the Upper Palaeolithic sites were linked to the remains of modern humans. Upper Palaeolithic stone tools were made from long, delicately shaped flakes, which provided fine scrapers, knives and spear points, but most importantly, the Upper Palaeolithic sites contained elaborate burials and magnificent rock art. The Upper Palaeolithic human remains were referred to as Cro-Magnon after a site that had provided several nearly complete skeletons, and the associated complex tool technology was given the name Aurignacian. The European archaeological sites showed a third stone tool technology, which was more difficult to understand. It also used prepared cores to strike stone flakes, but these were of poorer quality than the Aurignacian. By the 1880s scientists recognised that these stone tools were associated with Neanderthal-like skeletons and large Ice Age fauna such as mammoth and rhinoceros. They fell in time between the Lower and Upper Palaeolithic – hence the use of Middle Palaeolithic to define them. The name Mousterian was coined when the rich French Middle Palaeolithic site at Le Moustier was published in 1908 (Penniman 1952, 266).

Péringuey claimed that his hand-axes were as old as the most ancient in Europe, a revolutionary idea at the time (Deacon and Deacon 1999, 5). He is credited with moving away from European names for African archaeological sequences and, although his efforts provided a foundation for future research, he really only differentiated between 'ancient', referring to his hand-axes and cleavers, and 'modern', in the sense of small, finely made tools and rock art associated with the cave sites on the South African coast (Goodwin 1935, 324). Most importantly, he acknowledged that the more recent archaeology could be directly connected to the living and historical San of the region (Deacon 1990, 42). His ideas were briefly laid out in a precis he prepared in 1905 and then more fully six years later (Péringuey 1905, 1911).

Péringuey's attempts to separate the terminology for South African archaeology from its European counterpart would trigger a serious discussion of the issue in the 1920s. The terminology was refined by John Goodwin and Clarence van Riet Lowe in their 1929 monograph. Goodwin and Van Riet Lowe created three new names – Early Stone Age, Middle Stone Age and Later Stone Age – to accommodate and adapt the European Lower, Middle and Upper Palaeolithic to South African discoveries (Gowlett 1990, 19). Goodwin and Van Riet Lowe included Péringuey's hand-axes in the Early Stone Age, but initially were unable to associate specific human remains with the technology. They could find no unambiguous evidence of Neanderthals at any South African site. So, although their Middle Stone Age conformed to the European Middle Palaeolithic, they would not use European site names to define the African equivalents. The association of the Later Stone Age with the Upper Palaeolithic also created some mismatches. The long, beautifully shaped blades so typical of Aurignacian sites were extremely rare in South Africa. However, the microlithic tools were easily as complex and well made, and were connected to polished bone implements, decorations and rock art similar to the Upper Palaeolithic sites of Europe. The South African sites were not Aurignacian, but the similarity was strong enough that the term would be used to draw positive comparisons between the first modern Europeans and the Khoesan of South Africa at least until the 1950s.

Péringuey's association of the living San with the recent archaeological finds in South Africa led directly to his becoming a collector of human skeletons. He cultivated a network of amateurs who provided him with archaeological skeletons even before the museum had a formal Department of Archaeology (Morris 1992a, 4), but the most intense period of his collection was after the meeting of the South African Association for the Advancement of Science (S2A3) in Cape Town in August 1905.

The end of the South African War (the Boer War) in 1902 brought a period of uncertainty in which began a major restructuring of South

African society. Education and science had been on separate pathways in the republics and the colonies, but the negotiations from 1905 about a Union of South Africa opened the door for a unified policy to be modelled on Great Britain. Although people of colour were to be excluded from this process, there was a conscious effort to try to unify Boer and Brit in this new dispensation. Science was to take a special place in this restructuring.

The idea of S2A3 seems to have taken seed in 1900 from a modest suggestion by Sir David Gill, Her Majesty's astronomer at the Cape. Gill used his visit to England in May 1900 to suggest to the British Association that they hold one of their meetings in South Africa. The plan from his perspective at the Cape was to use the meeting to inaugurate a South African association modelled on the British one, in which non-academic scientists would also be welcome, as opposed to the South African Philosophical Society, which restricted member-ship to those with an academic qualification (Summers 1975, 86).

The British responded positively to the suggestion and in March 1901 the idea of the S2A3 was officially proposed by the engineer Theodore Reunert. In September of that same year, a Cape Town committee, with Gill as a member, approved the formation of the S2A3. The new S2A3 was one of the first post-war national institutions to be formed in South Africa. The first meeting of the S2A3 took place in Cape Town in 1903, but, as an impetus to the development of science in the Cape Colony, the South Africans requested that the parent British Association have a joint meeting with its new South African counterpart as soon as pos-sible. The invitation was accepted and in 1905 the British Association scheduled its 75th meeting for Cape Town in combination with the South African Association's third meeting (Morris 2002, 336).

The South Africans also invited other international organisations to participate in this joint meeting. Some 50 foreign (non-British) scientists were invited and those that accepted came from Canada, the United States, Denmark, Austria-Hungary, the Netherlands, Germany and even Japan (Morris 2002, 337).

Both the South Africans and the British felt that there was much to be gained by the collaboration. On the South African side, the 1905 meeting was a red-carpet event and a chance for the South African colonies to show off their scientific assets, but also an event that would hopefully stimulate science in all the southern African colonies. The joint meetings in Canada of the British and Canadian Associations of Science in 1884 and 1897 had been wildly successful and had resulted in substantial growth in scientific teaching and research in that country. For the British Association, the opportunity to visit South Africa was important not only because it was an exotic location for a meeting, but also because it was seen as an opportunity to expand its imperialistic ideals. Industrial development in the United Kingdom had levelled off and started to decline in the 1880s, and many of Britain's ruling elite (both political and scientific) began to see the Empire as a possible solution to their social and economic problems at home (Worboys 1981, 173). The British state saw that science could supply new technologies to expand industry and the Empire could provide economic opportunities as markets and sources of raw material, but only if its people and geography were better understood. In South Africa's case, there was a special need for unity. The devastating South African War (1899–1902) had left deep scars among the populace. Both the new S2A3 and the older British Association hoped that the joint meeting would create a camaraderie of science that would spark sympathetic feeling between the subjects of the United Kingdom and South Africa and a unity between Boer and Brit in the post-war reconstruction.

From an anthropological perspective, two visiting anthropologists of international note would leave lasting influences on the developing anthropological scene in South Africa. The chair of the anthropology section was the Cambridge academic Alfred Cort Haddon, but among the foreign delegates (that is to say, non-British) was Felix von Luschan, the director of the Königliches Museum für Volkerkunde (Ethnological Museum) in Berlin. He was as respected an anthropologist in the German-speaking world as Haddon was in the English-speaking world.

Haddon presented the chair's address for the anthropology section. The bulk of his paper was a state-of-knowledge address about the peopling of southern Africa, but the last three pages were a plea to learn more about the living people of the subcontinent. He was concerned that many of these people were vanishing and argued that it was up to science to gather data on them before they were gone. He implored researchers to study first those aspects that were being lost. He noted that anthropometrical data were everywhere wanting and comparative physiology was unstudied. Most important to Haddon was the need to study the psychology of these populations. Haddon's plea was loaded by the perceived need to use anthropological data to train administrators and he stressed that good administrators of colonies needed to understand the peoples they governed.

Von Luschan also used the 1905 meeting to present his ideas about the peopling of southern Africa. In particular, he pushed the Hamitic theory for the origin of the 'Hottentots'. He used physical, cultural and linguistic evidence to identify a fundamental separation of 'Bushmen' from 'Hottentots' (Von Luschan 1906, 1907). He, like Haddon, spoke of the need for research on the psychological characteristics of the people, along with much-needed new anthropometric data on southern African peoples, including body casts of living individuals (Morris 2002, 338). Von Luschan strongly supported the link between behaviour and race, and one of his strongest reasons for separating 'Bushmen' from 'Hottentots' was the perceived fact that the 'Hottentots' were shepherds because of their Hamitic ancestry. Although Haddon and Von Luschan shared the idea of Hamitic traits among the 'Hottentots' (Haddon 1909, 37), it was Von Luschan and the German ethnologists who would make the strongest link between behaviour and biology. The German ethnos theory assumed that nations (*volk*) each had their own culture, which stemmed from its group origin (Sharp 1981, 19). This concept of the union of race and culture was a critical part of German ethnological training and was especially strong among the missionary anthropologists who studied the peoples of German South West Africa (Namibia).

The link between race and culture would flow from the German ethnos theory into the developing Afrikaans-language anthropology in South Africa with political consequences in the development of apartheid (Sharp 1981, 32; Gordon 1988, 536), but it also fitted well in the colonial system. It was believed that understanding the race of people exposed their psychology and therefore made them easier to govern. Von Luschan emphasised the importance of Cape Town as a place to study the mix of races and hoped 'that one of our English colleagues or one of the colonial scientists will soon deal with these questions' (Von Luschan 1907, 8).

The immediate outcome of the 1905 joint meeting was that the S2A3 set up a standing committee for anthropology. The committee survived for only two years and had great difficulty getting co-operation from the various colonial governments. Despite this, the committee did have the effect of stimulating papers on anthropology at subsequent annual conferences (Shaw 1978, 5).

The 1905 meeting had a special significance for the developing field of physical anthropology. No academic departments of anatomy or anthropology existed in the South Africa of 1905, so the influence of the British Association on their development must at best be considered indirect, but not so for the museums. Haddon had made it a special interest to examine the collections of these institutions and he made specific recommendations about developing the quality of the ethnological collections and their accession records.

Péringuey was part of the organising committee for the joint meeting and was certainly in the audience for both Haddon's and Von Luschan's presentations. He almost certainly had private conversations with both of these men during their stay in Cape Town. Von Luschan persuaded Péringuey to make casts of San men from the Breakwater

Prison, recommending that accurate casts of live individuals and skeletons of dead ones would be the best way to make a record of these people. Péringuey agreed with Haddon and Von Luschan that the San as a people were on the verge of extinction and would soon follow their archaeological ancestors into the mists of time. He became convinced that the best way to make a record of these 'dying races' was to make casts

Figure 1.1: Staff of the South African Museum in 1920. Front row, left to right: Robert Lightfoot, Keppel Barnard, Louis Péringuey and Star Garabedian. Back row, left to right: Stanley Gilman, James Drury, Robert Tucker and Sidney Haughton. (Source: Summers 1975; courtesy of Iziko Museum)

of as many people as possible and to collect skeletons (Shaw 1988, 3). Although the first casts produced were not very successful, James Drury, the South African Museum's taxidermist, persevered with the casting programme, producing his first successful effort in 1907. He continued to produce casts of new individuals until Péringuey's death in 1924 (Summers 1975, 103; Booth 1988, 23; Davison 1993, 171). At the same time Péringuey began his most intense period of gathering skeletons through his network of amateur collectors and by sending Drury out into the field to excavate human remains. The untrained but extremely thorough Drury was a willing field excavator who eventually dug two sites of note – the Coldstream Cave and much later the cemetery at Colesberg – both of which produced large numbers of human skeletons.

Péringuey's skeleton-collecting activities have been well described, especially for the period after 1905 (Morris 1992a, 4; Legassick and Rassool 2000, 5). The initial specimens were brought to the museum in the 1890s as discoveries from disturbed archaeological contexts – most of which seem to have come from coastal exposures or amateur explorations of caves (Morris 1992a, 4). But the 1905 meeting ushered in Péringuey's most intensive collecting. Before 1905, he had occasionally used museum funds to cover the expenses his amateur collaborators incurred in sending specimens to him, but after 1905 he used the funds to regularly purchase human remains from people whose sole purpose was to gather human skeletons for profit. The nature of these new specimens was very different as they were frequently individuals who had been known in life and whose graves were exhumed for the sale of their contents (Morris 1987, 18; Legassick and Rassool 2000, 31). In 1905 Péringuey had 24 complete crania and many more fragmentary remains, but this had risen to 62 complete specimens by 1911 and 163 complete skulls by 1917 (Legassick and Rassool 2000, 6). The sample was a mix of 'ethnically identified' racial type specimens obtained from individuals known in life and archaeological skeletons, mostly from rock shelters along the south Cape coast. Péringuey had purchased three skeletons from the landowner of Coldstream Cave and then sent Drury to excavate in 1911. Drury recovered another 28 skeletons (Wilson and Van Rijssen 1990, 7) and nearly double that after Péringuey's death from the 'Bushman cemetery' outside of the town of Beaufort West in 1926 (Morris 1992b, 135).

Although he was busy collecting skeletons, Péringuey was not equipped to study them. There would be no anatomist/physical anthropologist in Cape Town until 1911 when the Medical School at the University of Cape Town was launched (Morris 2012, S154). In 1905 Péringuey wrote to Frank Shrubsall in England to find a scientific collaborator to study the human remains he was collecting.

Unfortunately, the very first letter that Péringuey wrote to Shrubsall has not survived, but Shrubsall's response of 5 November 1905 indicates that it must have been written sometime in October of that year.[1] Péringuey undoubtedly knew that Shrubsall had written a paper on the Khoesan human remains in European collections in 1898, but the direct impetus to write may have come from Péringuey's contact with Haddon at the meeting in Cape Town in August. Haddon had been a part-time lecturer in physical anthropology at Cambridge at the exact time that Shrubsall was a student there (Spencer 1997a, 469) and it seems very likely that he had recommended Shrubsall to Péringuey as a collaborator.

FRANK C. SHRUBSALL (1874–1935)

Shrubsall, the son of a master mariner, was a scholarship student who went to Clare College, Cambridge, in 1892 to study the Natural Science Tripos. He received his Bachelor of Arts in 1895, Master's degree in 1899, MBChB (Bachelor of Medicine, Bachelor of Surgery) in 1901 and his Doctor of Medicine in 1903. He obtained a first in both parts of the Natural Science Tripos with a specific interest in anatomy and craniology.[2] His years at Cambridge overlapped with a fellow student, Wynfrid L.H. Duckworth at Jesus College, and the two would have been studying the crania in the collection of the Department of Anatomy at the same time. Haddon was a part-time lecturer in the department from 1894 to 1898, and Duckworth replaced him as the first full-time lecturer in physical anthropology from 1898 (Goldby 1956). Shrubsall focused on African skeletons for his Master's research, starting first with specimens from Tenerife in the Canary Islands (Shrubsall 1896), then to southern Africa (Shrubsall 1898a, 1898c) and then elsewhere in Africa (Shrubsall 1898b, 1901, 1902). All of his research was conducted on specimens held primarily at Cambridge or at the Royal College of Surgeons in London. Once he had graduated from medicine, Shrubsall took the post of resident medical officer at the Hospital for Consumption and Diseases of the Chest in London (Hamer 1935, 646). Although he maintained an interest in anthropology and was a member of the Royal Anthropological Society and the anthropology section of the British Association for the Advancement of Science, his hospital duties drew him into clinical medicine and he was much less engaged in anthropology by late 1905 when he received Péringuey's letter.

Shrubsall's response to Péringuey's invitation to work on the skulls from the South African Museum was one of great excitement. Péringuey must have sent him a list of specimens because he immediately responded with the catalogue numbers that interested him. He ends his letter: 'The matter of keenest interest to me is to trace the track of the "Bush" from equatorial Africa to the Cape, which I hope to do someday by the aid of more specimens and newer methods of investigation.'[3] Within a year of first contact, Péringuey had sent 24 crania to London, 10 of which were complete enough for Shrubsall to begin his analysis. In a letter dated 8 October 1906, Shrubsall remarks that he was gathering data on a range of Egyptian crania in order to test Von Luschan's Hamitic hypothesis. Along with this letter he forwarded a draft manuscript to Péringuey, but the warning was already there that he was struggling to find time for research. He commented: 'I regret taking so long but I have much office work as a medical super-intendent beside professional duties and some lecturing so my time to get clear for measuring is rather scanty.'[4]

By the post of 22 March 1907, Shrubsall had some exciting news for Péringuey. He had presented a paper to the Royal Anthropological Society based on Peringuey's specimens and there had been an active and engaging response from the audience. He had concluded that all of the specimens from the cave sites were broadly 'Negroid', but did not resemble the 'Negroids' of Grimaldi caves. They were very similar to both the Kalahari 'Bushmen' at the College of Surgeons and the various 'Hottentot' skeletons he had accessed in the European collections. Arthur Keith was a key discussant at the meeting and Shrubsall summarised for Péringuey Keith's view 'that the evidence showed the age of the Strandloopers to be reasonably prior to the advent of the European, Hottentot and Bantu but how far anterior not in any way proven. He does not regard the Negro and still less the Bushman as a primitive people but as a highly evolved race characterised by infantilism.'[5] Shrubsall continued in his report to say that Keith felt the short stature of the 'Bushmen' was not the result of their being related to some kind of 'pygmy race', but rather that their 'uncertain and insufficient food supplies produced a dwarfing effect'. The balance of Shrubsall's letter discusses details for publication and a final warning: 'I do not see that the paper can be ready till the summer as I have always calls on my evenings for official work and practically Sunday mornings at home and a few Saturday afternoons at museums represents my whole available time.'[6]

However, Shrubsall did manage to produce a publishable manu-script for the *Annals of the South African Museum* in the latter part of 1907 (Shrubsall 1907). In all, he presented data on 43 'Bushmen' and 30 'Hottentots', along with the cave dweller skeletons that Péringuey had sent from Cape Town. His conclusions were that 'Strandlooper-Bush-Hottentots formed a single group with great antiquity probably throughout Africa'.[7] For Shrubsall, 'Hottentots' seem to have been intermediate in character between the 'Bush' and the 'Bantu'. The article also contained a section on the post-cranial bones of the archaeological skeletons. Although this was a small sample, it marks the first note in the literature about the post-cranial variation of Khoesan peoples.

We have no surviving correspondence between Péringuey and Shrubsall for the next four years, but the two continued to collaborate and Shrubsall produced a smaller paper based on a larger sample to include as an addendum to Péringuey's 1911 opus. The updated paper included 62 more specimens (Shrubsall 1911). The earlier correspondence suggests that Péringuey was sending specimens on to Shrubsall as they became available and that Shrubsall seldom kept them for more than a year or two. But then in 1911 Shrubsall wrote a brief letter to Péringuey to enquire about either borrowing a mineralised skull from a consolidated midden that Keith had expressed interest in or to have Péringuey provide some technical photographs of the specimen. Not only does this tell us that Shrubsall was still discussing Khoesan origins with Keith, but it also makes it clear that he was aware that other researchers were looking at the same material. He notes briefly that 'Dr Pöch has 76 skeletons to our 9'.[8] The low number of his skeletons suggests that he is referring to post-cranial remains and that he is speaking of the collection of recent San skeletons that the Austrian professor had gathered under dubious circumstances in the southern Kalahari (Legassick and Rassool 2000, 15). In the end, Rudolf Pöch only produced a brief note on one skull that he saw in Pretoria (Pöch 1909) and he never described the large series he collected in Bechuanaland (now Botswana).

After 1911, the correspondence and the publications again are lost until 10 September 1918, when Shrubsall wrote to Péringuey after a long hiatus, apologising for his silence and reporting on his activities over the war years. He had volunteered for service, but his health was poor and instead he joined the medical anti-aircraft patrols on the nights of raids. He also told Péringuey that his medical duties were becoming heavier.[9] In fact, Shrubsall had moved into what we would recognise today as public health and he was engaged in designing a health system for the London County Council on the treatment of 'mentally deficient' children and special needs schools (Hamer 1935, 646). The reason Shrubsall was writing was to try one more time to complete the large-scale project that had originally been proposed between them. Despite the fact that he was very busy, he felt honour bound to complete the task. In his responding letter Péringuey acknowledged the sacrifice that many had made over the war years, adding: 'I cannot help feeling from the tenor of your letter that you are eager still to continue the work on the material I have been at some pains to bring together. At present there is no one here who is able to devote his whole time to craniology.'[10]

If Péringuey was expecting a rapid response and a revitalisation of the work, he was to be disappointed. With no word from Shrubsall, Péringuey finally wrote to him on 3 January 1921 with one last offer. After telling Shrubsall that the Medical School in Cape Town was now operational, he complained that although the professor of anatomy (Robert B. Thomson) had produced one paper (Thomson 1913), he had since left academia and there was no one locally to look at the skeletons. With many new specimens to include, Péringuey lamented: 'Who is going to work out and co-ordinate the results of my finds, mensuration etc. if you abandon the task which you had so willingly undertaken, and which I am so anxious to see continuing on its way.'[11] He asked Shrubsall to at least submit a palliative, summarising the results based on measurements, which he would then ask his museum assistants to compile. Shrubsall agreed to writing a summary based on submitted data, saying: 'I am really awfully sorry and know even now I am a broken reed but I like you can find no one else but I will promise to tackle the forms if not too many and to endeavour if I clear my immediate duties in time to have a fresh look at old work.'[12]

The final paper (Shrubsall 1922, 185) is indeed a shadow of the earlier works, but it does provide some very useful data. The total sample is 207 skulls, the majority archaeological, and they are sourced not only from the South African Museum, but also the Albany Museum in Grahamstown and the McGregor Museum in Kimberley. Shrubsall did not see any of the new material first-hand, which meant that the paper was based on statistics rather than description. Péringuey divided the sample into four groups: his ethnographically identified samples of Kalahari 'Bush' (including some exhumations of recently deceased people) and 'Hottentots'; and two sets of archaeological remains, one from the Cape or Southern 'Bush' (the inland of the Cape Province) and the other from the coastal caves and shell middens.

Péringuey wrote one last letter on 29 April 1921 in which he told Shrubsall of a promising student who was keen to continue the research, but had subsequently left South Africa and once again he was unable to find someone to continue. He added: 'The greatest for me now is to find a worker that is able to do justice to this fascinating subject.'[13]

EUGÈNE PITTARD (1867–1962)

Just less than a year after his last correspondence with Shrubsall, Péringuey received an unexpected letter from Eugène Pittard in Geneva, Switzerland. The content of the letter confirms that Péringuey had known Pittard before the war, but the letter gives no details about the level of their acquaintance.[14] Pittard noted that two events had prompted him to renew their contact: the recent announcement of new discoveries in South Africa (not mentioned, but most likely to be the Boskop skull) and the visit of a missionary to his office who showed him some interesting tools in flint from South Africa. Pittard enclosed some of his recent publications and requested the same from Péringuey, but then went on to propose a much larger exchange. In return for some European skulls from the Alps and some Mousterian and Aurignacian tools, Pittard requested some human skulls and Palaeolithic tools from South Africa. We do not have Péringuey's immediate response, but it is clear from the next letter from Pittard that Péringuey made a significant proposal.[15] Pittard responded that he would be very happy to receive the collection of 'Hottentot' skeletons and skulls and that he would like to produce a monograph on them similar to his book on the Valais skulls from Switzerland (Pittard 1910).

Pittard was not an Africanist. His interests were overwhelmingly focused on Europe, in particular the ancient Mousterian and Aurignacian peoples of Western Europe, and the living peoples of the Balkans. Early in his career he had studied at Paul Broca's anthropology laboratory at the École des Hautes Études in Paris and had become a disciple of Léonce Manouvrier, switching from zoology to anthropology (Popescu-Spineni 2015, 157). He returned to his native Switzerland and completed his doctorate in 1899 on the craniology of the people of the Rhone Valley, published as a monograph in 1910. Shortly after graduating, Pittard travelled to Romania at the invitation of Prince George Bibescu and began what would be three decades of research on the peoples of the Balkans and Asia Minor. He was particularly struck by the ethnic complexity of the region, finding 'Romanians, Tartars, Turkish, Armenians, Bulgarians, Kurdish, Gypsies, [and] Jewish' all living together (Popescu-Spineni 2015, 158). Pittard's Balkan research was paralleled by his continued interest in prehistory. In 1905 he excavated cave sites in the Dordogne in France where he worked on the typological assessment of stone and bone tools from earlier Mousterian and later Aurignacian sites. Pittard had a very broad view of anthropology and his extensive writings include a mix of skeletal biology, culture and ethnic studies. His broad interests made him out of step with many of the other European anthropologists of the day. He was deeply concerned about the ideology of race and nationalism and published an essay on the topic at the height of the First World War (Pittard 1916). He followed this in 1924 with an important volume summarising his ideas on race, titled *Les races et l'histoire*. He argued that race (a group of individuals that he defined as being related to each other by the blood of their parents) had nothing in common with nationality, language or morals: '*La race est un fait biologique, la langue est un fait social*' (Race is a biological fact, language is a social construct) (Reubi 2010, 41). Pittard's Balkan experience prevented him from believing that there were any pure races in Europe and he believed that the mixing of peoples enriched human groups and prevented their degeneration. He assumed that his own classification of peoples was of practical utility only and there were no such things as inferior or superior races (Popescu-Spineni 2015, 160).

So, the person that Péringuey was about to entrust with his entire collection of South African skeletons was a skilled racial typologist who did not believe in race and had no real experience in the study of African populations. But Pittard had something in Geneva that Shrubsall did not have in London – a fully functioning laboratory of anthropology. Pittard also now had students and younger researchers working with him who could do a lot of the work and hopefully produce many more publications than Shrubsall had been able to.

Pittard's acceptance of Péringuey's request to analyse the skeletons resulted in a series of letters to clarify exactly what was to be done. On 13 July 1922 Péringuey was very clear that none of the human remains could be transferred permanently to Geneva, but a long-term loan would certainly be possible.[16] Any plans for publication would need to be in English. In his response Pittard asked to delay the start of the project for a little while as he was deeply involved in a project that needed to be completed first.[17] This was the production of his important book on the history of races (Pittard 1924), the manuscript of which was nearing completion. He also indicated that the work on the Khoesan could be published either in French or in English. Péringuey wrote back that although Pittard would have to make the decision about publication, it would be important to publish in English as that was the working scientific language in South Africa, even above the local Dutch, which was also a state language.[18] Péringuey then asked Pittard to give him advance notice when he was ready to begin the research.

After waiting almost ten months for Pittard to send him notice that he was ready to start, on 23 July Péringuey finally wrote directly to Pittard asking if he was ready.[19] Péringuey was obviously under some pressure. He wrote that he had received criticism that he was not offering the task to a local scientist when there were already a number of South African professors interested in prehistory. Things had changed a lot since the situation with Shrubsall in the decade before. In fact, two medical schools had launched, one in 1911 at the South African College in Cape Town (the University of Cape Town after 1918) and the other in 1919 at the University College, Johannesburg (the University of the Witwatersrand after 1922) (Morris 2012). At both schools the first incumbent professors did little research, but the next two, especially Raymond Dart in Johannesburg, were keen to do anthropological research. It is clear from his letter of 23 July that Péringuey was holding them off to fulfil his promise to Pittard. To Péringuey's great relief, Pittard wrote back on 11 September agreeing to start work.[20] He laid out a programme in which he would take on the craniology and the details of the vertebral columns himself, but he would arrange for colleagues and students to work on individual elements. He would take on the task of assembling all of the data into one single monograph that he would author. Pittard's methodology would mean that most of the papers would not consider individual skeletons, but would deal with each osteological element as if it were unconnected from the other bones of the body. This was not an unusual practice at the time as racial variation was believed to be as evident in each bony element as it was in the individual as a whole. Péringuey was overjoyed and immediately arranged for seven cases of bones to be sent to Geneva.[21] He included a long passage in his letter outlining his views on the peopling of southern Africa. On 30 November Péringuey wrote that he had sent the third consignment of bones, this time in two boxes, and then on 10 January 1924 he sent a final consignment of five skulls.[22]

Pittard wrote back to Péringuey on 6 February, acknowledging that the first of the cases had arrived.[23] He had already spoken to some of his graduated students and they were interested in getting involved. Pittard reaffirmed that he would take on the role of co-ordinator and would be responsible for summarising the work in a single volume, but he made it clear to Péringuey that he did not speak English well enough to do the translation and that this would have to be Péringuey's task. The short preliminary papers would be in French journals. This was to be the last letter that Péringuey received. He died at the age of 69 walking home from his museum office on 20 February 1924 (Summers 1975, 121).

The death of Péringuey would affect the project in two very important ways. Pittard's study of the skeletons from the South African Museum was structured around a classification of South African peoples set up by Péringuey in his letter of 25 October 1923. With Péringuey's death, there would be no room to modify or discuss the classification and Pittard would accept it without change as the basis for the analysis. In the letter, Péringuey outlined his belief that the 'Strandloopers', either from caves or sand dunes, represented an early population that was not 'Bushmen', but had migrated earlier to South Africa. Péringuey had suggested they were of the same level as the Aurignacian people of Europe. The 'Strandloopers', the 'Bushmen' and the 'Hottentot' each represented a wave of immigration reaching South Africa at different times. Péringuey divided the sample of skeletons in a similar manner to what he had done for Shrubsall, but with even more detail, and noted these six groups in his letter:

1) Strandloopers from Rock Shelters
2) Strandloopers from Sand Dunes near Kitchen middens.
3) Bushmen from the interior of the Cape Colony south of the Orange River
4) Bushmen from the Kalahari and Southwest Africa.
5) Hottentots: a small group of rare skulls mostly from Little Namaqualand.
6) The Korana (subsequently called Griqua): a mixed Hottentot-Bantu group from along the Orange River.[24]

The last group, the Griqua, was a new addition, examples of which had not been sent to Shrubsall for analysis. It consisted of about twenty skeletons, which had been excavated from old graves in the Kimberley area by Vernon Brink between 1918 and 1920. Brink, an interested amateur, had basically gone grave robbing, asking farmers in the area if there were Griqua graves on their farms and then digging up the skeletons for his study (Brink 1923). Brink planned on doing a major description of the bones while he studied medicine in Oxford, and although he had taken all of the specimens he excavated with him to the United Kingdom, he decided in the end to send the bones to the South African Museum in April 1923 when he realised he could not find the time to do the task. In his letter of thanks to Brink, Péringuey noted that he now had the promise of one of the leading European physical anthropologists to examine the specimens.[25] Although some views on the origin of the Khoesan did change in the decade before the Second World War, Pittard stuck to Péringuey's outline and there were no further refinements to the model.

The second impact of Péringuey's death was that Pittard had lost his translator for the upcoming works. Although Pittard continued to talk about the possibility of completing a monograph on the skeletons, ultimately it was only the set of preliminary reports, all in French, which were published.[26]

PITTARD'S PAPERS

Pittard's Cape Town benefactor had gone, but he did follow through with his promise to work on the skeletons. Between 1926 and 1950, he and his students produced 53 papers,[27] which discussed various aspects of the skeletons: 32 papers on craniology or cranial anatomy; 9 on the jaws and teeth; and 12 on post-cranial elements. As he had indicated in his correspondence with Péringuey, Pittard's first paper was on craniology (Pittard 1925/26), prepared for a 70th birthday volume in honour of the Croatian palaeoanthropologist, Karl Gorjanović-Kramberger. Most of the first part of the paper recaps the state of knowledge at the time about living or historical Khoesan peoples. Pittard quotes descriptions of the physical features of these people, as seen by European explorers across the African continent. He then draws these living Khoesan peoples into the past by introducing Péringuey's classification system and the specimens provided in each category – 139 crania complete enough to analyse.[28]

The rest of the analysis is not particularly useful or perceptive in terms of our knowledge today. With a focus only on cranial shape calculated as the cranial index, the discussion is concerned only with the ratio of dolichocephalic (long-headed), mesocephalic (medium-headed) and brachycephalic (short-headed) individuals in each population. Pittard did not use sophisticated mathematics in his research and the focus on cranial shape in Pittard's work was pure nineteenth-century craniology without any statistical elegance. His conclusion was that for the most part 'Bushman-Hottentots' were long-headed, but there were signs of some genetic mixing with short-headed people. Pittard was relying on the nineteenth-century approach to typology where only one or two key measurements were used to define differences between individuals and populations. This is useless information from a modern perspective and even at the time it provided weaker evidence for population relationships than Shrubsall's more mathematical approach, which required multiple measurements and attempted to provide a summary of differences.

What followed over the next few years was a series of papers describing specific craniological features. Some, such as Flower's dental index (Pittard and Baïcoyano 1928) and the frequency of Carabelli's trait (Périer 1930), are useful in describing the size and shape of teeth and there are some interesting evolutionary relationships in these features, but many were simply thorough (and pointless) craniological calculations without any understanding of the reason for shape differences in the skull. The post-cranial papers had more value as there was a more functional approach to the anatomical detail, but some of them were concerned with simple mathematical indices that describe the flattening of bone shafts, but kept their discussion to racial rather than functional explanations. Perhaps the most interesting paper was one that looked at long bone lengths for the reconstruction of height (Pittard and Comas 1930), something that remains a point of interest for modern anthropologists. The fact that many of the papers were co-authored with students or with recent graduates is more important than the actual content of the papers. Pittard was using the skeleton collection to train students in anthropological technique. At least one of them, Juan Comas, moved to Mexico and was to become one of the most important physical anthropologists in Latin America (Vargas 1997, 291). Pittard's meticulous training of Comas provided him with detailed anatomical knowledge of the skeleton and its measurement, but his more thorough understanding of statistical methods helped him to reconstruct the biological history of the peoples of Mesoamerica. Despite, or perhaps because of his typological training, Comas was critical of racial studies and was a vocal critic of the old anthropometric techniques and the fallacy of racism (Vargas 1997, 291). Many of Pittard's students came from the Balkans and Turkey, where Pittard's writing of ethnographic history had made him both well known and very well respected (Popescu-Spineni 2015, 163).

Although the total number of papers is impressive, none was published outside of Europe and all of them were in French. Most of the papers were short notes, or preliminary papers as Pittard wished to call them. These were in the local *Bulletin de la Société suisse d'anthropologie et d'ethnologie*, the *Archives suisses d'anthropologie générale*, a journal launched by Pittard himself in 1914 as the publication arm of the Swiss Institute of Anthropology, and the *Archives des sciences*, edited by the Société de physique et d'histoire naturelle de Genève. Several papers were submitted to prestigious French journals, such as *L'Anthropologie* and the *Revue anthropologique*. At least one thesis was published on a topic linked to the skeletons from South Africa (Grintzesco 1933). Péringuey had asked that some papers be published in the *Annals of the South African Museum*, but in the end, no papers were ever submitted either in French or in English.

THE SKELETONS' HOMEWARD JOURNEY

Upon becoming the new director of the South African Museum in Cape Town after Péringuey's death, one of Leonard Gill's earliest tasks was to follow up on loans of material to various researchers. He wrote to Pittard at the beginning of 1925 to enquire about how the research was progressing and when the project would likely be finished.[29] A series of letters followed, sorting out details of catalogue numbers and information about archaeological context, but by June of the following year there was a warning that Gill was being put under some pressure to have the skeletons returned to South Africa and that at least one of the museum trustees felt that Péringuey had been wrong to send them out of the country.[30] The Cape Town archaeologist John Goodwin passed through Geneva in 1930 and made a courtesy call on Pittard. He wrote to Gill afterwards, saying: 'I have just been seeing Pittard and *our* skulls. They are all safe and a good part of the material is ready for writing up for publication.'[31] After several letters about the possibility of publishing in South Africa, Gill told Pittard directly that there has been 'a good deal of reproach for letting it [the collection of skeletons] remain so long inaccessible to students of anthropology in South Africa.'[32] Pittard responded by saying that he too was concerned at the length of time the study was taking and that he had abandoned other research to focus on the African material.[33] However, he was reluctant to return the skeletons until he had achieved the work that he had engaged to do with Péringuey. Finally, on 13 May 1938, Gill and the trustees of the South African Museum wrote a formal request for the return of the skeletons.[34] Pittard agreed to return the cranial material first, followed a few months later by the post-cranial remains to give him and his team the time to finish the post-cranial work.[35] By then the skeletons had been in Geneva for fourteen years.

The final process of packing up the skeletons for their return had begun. Mr H.T. Andrews of the South African delegation to the League of Nations helped to smooth the way through some of the regulations and five cases left Marseilles for South Africa on 7 February 1939. But not everything was going smoothly. Gill was extremely concerned about the possibility of war and the bones were travelling slowly down the west coast of Africa, scheduled to arrive in Cape Town only in April.[36] He urged Pittard to pack up the rest of the material as soon as possible. Then came bad news: Pittard wrote to Gill that there had been a delay in research because the laboratory was being moved.[37] The bones were packed and safe, but Pittard was not prepared to send them on their journey. The international situation was just too dangerous.

The post-cranial skeletons and several mandibles that had not been sent back with the skulls spent the war years sealed in cases in the storeroom of the Department of Anthropology in Geneva. No further research on the bones took place. The fact of the war made any claims from South Africa pointless and the change of director at the South African Museum in 1942 meant the bones were almost forgotten. In late 1947, a planned research trip to Europe by Dr J.A. Keen of the Department of Anatomy at the University of Cape Town prompted Dr E.N. Keen (the son of J.A. Keen) to ask the new director of the museum, Dr Keppel Barnard, if he knew whether the sample was available for study in Geneva or if there were plans for the return of the material. With this spur, Keppel pulled the Pittard records from his files and began the final process to repatriate the skeletons.[38] Writing to the South African high commissioner in London, Barnard said: 'In the past Prof Pittard made one excuse after another to retain the material a little longer; but if, as we are given to understand, the material is packed ready for shipment, no further excuses should be possible. Any such proposal to retain the material any longer must, in fact, be firmly declined.'[39] Pittard, now very elderly and no longer interested in fulfilling his old promise to Péringuey, had no objection. The last of the Khoesan skeletons in Geneva began their journey back to Cape Town on 23 June 1948.

A final note about this episode concerns the state of the skeletons when they arrived. A few months after the boxes arrived, Barnard asked the museum ethnologist, Margaret Shaw, to unpack them, with the help of the museum technician, Charles Thorne. Upon opening the boxes, they discovered that Pittard had sorted the skeletons into individual bones and that each box contained not a skeleton, but *all* of the femora, *all* of the vertebra, and so on. The one saving grace was that Pittard had been meticulous in putting the museum registration number onto every single bone, so it was possible to 'reconstruct' the skeletons from the numbers. Shaw and Thorne 'spent the most of one very hot February rebuilding them', as Shaw recalled in an interview on 26 June 1991.[40]

NOTES

1   Iziko Museum, Cape Town, letter from Frank Shrubsall to Louis Péringuey, 5 November 1905.
2   Alexandra I. Browne, archivist at Clare College, Cambridge, letter to author, 28 May 2015.
3   Iziko Museum, Cape Town, letter from Shrubsall to Péringuey, 5 November 1905.
4   Iziko Museum, Cape Town, letter from Shrubsall to Péringuey, 8 October 1906.
5   Iziko Museum, Cape Town, letter from Shrubsall to Péringuey, 22 March 1907.
6   Iziko Museum, Cape Town, letter from Shrubsall to Péringuey, 22 March 1907.
7   The use of the term 'Strandlooper' (usually 'Strandloper'; literally, beach walker) is specific to many of the early writers although it has now fallen entirely out of use. Khoesan people obtaining a living on the seaside had been noted by Jan van Riebeeck in the mid-1600s and it was assumed that these were an entirely different people from the San hunters of the interior and the pastoralist Khoekhoen. Most historians and archaeologists today support the idea that Strandlopers were San or Khoekhoen who were seasonally making use of marine resources.
8   Iziko Museum, Cape Town, letter from Shrubsall to Péringuey, 24 March 1911.
9   Iziko Museum, Cape Town, letter from Shrubsall to Péringuey, 10 September 1918.
10 Iziko Museum, Cape Town, letter from Péringuey to Shrubsall, 27 October 1918.
11 Iziko Museum, Cape Town, letter from Péringuey to Shrubsall, 3 January 1921.
12 Iziko Museum, Cape Town, letter from Shrubsall to Péringuey, 7 March 1921.
13 Iziko Museum, Cape Town, letter from Péringuey to Shrubsall, 29 April 1921.
14 Iziko Museum, Cape Town, letter from Eugène Pittard to Louis Péringuey, 4 February 1922.
15 Iziko Museum, Cape Town, letter from Pittard to Péringuey, 9 June 1922.
16 Iziko Museum, Cape Town, letter from Péringuey to Pittard, 13 July 1922.
17 Iziko Museum, Cape Town, letter from Pittard to Péringuey, 6 September 1922.
18 Iziko Museum, Cape Town, letter from Péringuey to Pittard, 13 October 1922.
19 Iziko Museum, Cape Town, letter from Péringuey to Pittard, 23 July 1923.
20 Iziko Museum, Cape Town, letter from Pittard to Péringuey, 11 September 1923.
21 Iziko Museum, Cape Town, letter from Péringuey to Pittard, 25 October 1923.
22 Iziko Museum, Cape Town, letter from Péringuey to Pittard, 30 November 1923; Iziko Museum, Cape Town, letter from Péringuey to Pittard, 10 January 1924.
23 Iziko Museum, Cape Town, letter from Pittard to Péringuey, 6 February 1924.
24 Iziko Museum, Cape Town, letter from Péringuey to Pittard, 25 October 1923.
25 Iziko Museum, Cape Town, Louis Péringuey to Vernon Brink, 7 June 1923.
26 Iziko Museum, Cape Town, letter from Eugène Pittard to Leonard Gill, 6 March 1925; Iziko Museum, Cape Town, letter from Pittard to Gill, 11 February 1929; Iziko Museum, Cape Town, letter from Pittard to Gill, 22 March 1930.
27 A complete set of these papers is stored at the Social History Library, Iziko Museum, Cape Town.
28 The maximum number of specimens used in any of the analyses was from Pittard (1939) – 162 individuals.
29 Iziko Museum, Cape Town, letter from Gill to Pittard, 22 January 1925.

30 Iziko Museum, Cape Town, letter from Gill to Pittard, 16 June 1926.
31 Iziko Museum, Cape Town, letter from A.J.H. Goodwin to Leonard Gill, 14 January 1930.
32 Iziko Museum, Cape Town, letter from Gill to Pittard, 7 December 1934.
33 Iziko Museum, Cape Town, letter from Pittard to Gill, 9 January 1935.
34 Iziko Museum, Cape Town, letter from Gill to Pittard, 13 May 1938.
35 Iziko Museum, Cape Town, letter from Pittard to Gill, 14 June 1938.
36 Iziko Museum, Cape Town, letter from Gill to Pittard, 25 March 1939.
37 Iziko Museum, Cape Town, letter from Pittard to Gill, 21 July 1939.
38 Iziko Museum, Cape Town, letter from Keppel Barnard to E.N. Keen, 2 December 1947.
39 Iziko Museum, Cape Town, letter from Barnard to High Commission, 11 March 1948.
40 Interview with Margaret Shaw, 26 June 1991, South African Museum, Cape Town.

2

Boskop: The First South African Fossil Human Celebrity

The first two decades of the twentieth century were formative times for the field of palaeoanthropology. Many of the important fossil discoveries were still new. The site of Le Moustier in France, which provided the name of the stone tool culture associated with Neanderthals, was discovered in 1908, and in the same year the nearly complete skeleton of a Neanderthal was excavated in the cave at La Chapelle-aux-Saints, adding to the fragmentary remains that had been recovered before. The consensus at the time was that Neanderthals made up a stage of human evolution, more primitive than but ancestral to the Cro-Magnon people, whose skeletons had first been discovered in 1868. The Cro-Magnon people were the makers of Aurignacian tools, implements that were felt to be the beginning of the technological revolution leading to modern humans. The question of racial origins crept in with the discovery of human remains from Chancelade in France in 1888 and two skeletons in 1901 from the Italian site of Grimaldi. Together, these specimens were said to represent 'Caucasoid', 'Mongoloid' and 'Negroid' races, indicating that the origin of all modern peoples must have been in Europe – an anthropological Garden of Eden. Then, in 1912, the critical evidence of a missing link between apes and humans was found in Piltdown in England. Although ultimately the Piltdown skull was shown to be a forgery, its presence in the scientific literature at the base of the human evolutionary tree was virtually unchallenged for 30 years. All of these sites were in Europe and all that was really at issue was which part of Europe had been the place where humans had evolved – England, France, Germany or Italy. Evidence from elsewhere was dismissed as unimportant or not really about humans. The interpretation of the time was quite clear: humanity had evolved and diversified in Europe, the current home of the most advanced form of humanity.

None of these great anthropological debates was an issue in the mind of Piet Botha in October 1913 as he dug a drainage ditch on his farm Koloniesplaats, a few kilometres from the village of Boskop in what was then the Transvaal Province of South Africa. The shovels hit some darkly stained bones encased in a soil matrix at about 1.3 metres deep in the trench and about 80 metres from the bank of the Mooi River (Haughton, Thomson and Péringuey 1917, 2). At the time of the trench excavation, Botha was being visited by J.L. Groenewald, an old friend from Adelaide in the eastern part of the Cape Province. The two friends debated between themselves whether or not the bones were human and decided to ask a local medical man to identify them, wagering a shilling on the outcome.[1] Their local expert decided the bones were not human, but Groenewald was sceptical. Botha was going to throw the bones away, but Groenewald asked if he could take them with him when he went home and get another opinion from the nearby Port Elizabeth Museum. Botha did not object. Later that year Groenewald took the skullcap with him to show Frederick FitzSimons, the director of the museum in Port Elizabeth. FitzSimons immediately recognised that the skullcap was human, but that it was very old and therefore very important. In his opinion it would be worth exploring further with an excavation at the site. Groenewald wrote to Botha in Dutch requesting permission and Botha consented by telegram, also agreeing to the donation of the bones to the Port Elizabeth Museum. FitzSimons visited Boskop, met with Botha, and excavated at the site, recovering a few more fragments and some soil-encrusted post-cranial remains. Unfortunately, the site had become waterlogged as the drain was still incomplete and FitzSimons had to stop digging. FitzSimons told Botha that he would like to come back and dig again once the site was drained and Botha agreed. The delay was the start of what was to become a very challenging situation involving three museums.

At this point, an unknown person (presumably Botha himself) contacted the Transvaal Museum in Pretoria about the site. The museum wrote to FitzSimons telling him that Botha had actually intended to

Figure 2.1: The cranium of the Boskop specimen, as illustrated in 'Preliminary Note on the Ancient Human-Skull Remains from the Transvaal'. (Source: Haughton, Thomson and Péringuey 1917)

give the skull to them. FitzSimons asked for clarification and the telegraphic response said that the Transvaal Museum 'strongly objects to you interfering in our province. Overlapping on your side. Owner intended skull for Pretoria Museum. Person who handed same to you was not in lawful possession of it.'[2] The Port Elizabeth Museum board conferred with Groenewald about the original donation and, satisfying themselves that the events of discovery did not support Pretoria's story, made a decision to keep the skull and to send it to England for analysis. Louis Péringuey at the South African Museum now got involved by writing to the Port Elizabeth Museum board chairperson, complaining that sending the skull to England was unnecessary and that local scientists in Cape Town were available for the analysis.[3] In this Péringuey had clearly changed his tune from the decade before when he had sent his skeletons to Frank Shrubsall for analysis. Some quick negotiation between FitzSimons and Péringuey resulted in an agreement that Péringuey's team in Cape Town would analyse the skull and help to fund further excavations at the site. FitzSimons would accession all of the remains in Port Elizabeth. Without access to the skull itself, the Transvaal Museum fell out of the race.

In the meantime, Botha had become very difficult to deal with. FitzSimons wanted to continue the excavation, but Botha would not respond to letters from Port Elizabeth and FitzSimons had to act through an intermediary. Now Botha demanded financial compensation for any further work to be done on the farm. FitzSimons tried to downplay the importance of the discovery to keep the amount as low as he could. In a letter to Péringuey, he said: 'These Boer peasants are very cunning and Botha would no doubt have surmised the find was more important than I am giving him to believe.'[4] FitzSimons offered Botha £20 for access and a further £20 if any more finds were made, but Botha demurred. At one point Botha asked for £500, but FitzSimons rejected that outright. In the end, they agreed to a single payment of £50. Botha would provide housing for the excavators, but would charge a per diem fee for subsistence. The skull and all subsequent fossils would be the property of the Port Elizabeth Museum.[5] FitzSimons lamented in a telegram to Péringuey that 'it was impossible to get better terms after the intervention of the Transvaal Museum'.[6] To get an idea of the magnitude of this cost, data from the first decade of the twentieth century indicate that the monthly wage of a European labourer was about £9 and a skilled European craftsman perhaps about double that (De Zwart 2011, 60). Botha drove a hard bargain.

Now that the financial dealing was done, Sidney Haughton, a young geologist from the South African Museum, was sent to Boskop to re-excavate the site. In the meantime Botha had completed the drain and Haughton was able to do a thorough survey. He was very disappointed as he could find no other evidence. He noted that this was the fourth excavation on the site, as a team from the Transvaal Museum had dug as well, unknown to Péringuey and FitzSimons.[7] He lamented that FitzSimons had had the opportunity, but had not completed the original excavation. In a letter to Péringuey, it is clear that FitzSimons was very angry that the interference of the Transvaal Museum had not only driven up the costs, but had also resulted in a delay of ten months.[8] Péringuey wrote back to FitzSimons, commiserating about the underhanded nature of the dealings with the museum in Pretoria, but he recommended that FitzSimons drop the matter, as the Transvaal Museum people obviously did not find anything in their excavation.[9]

The analysis of the skeletal remains continued for about a year after Haughton's return from the site. An endocranial cast of the brain was sent overseas for analysis by the anatomist Grafton Elliot Smith, then in Manchester, but the accounts of the geology, archaeology and osteology were done by local researchers. In the published report, Haughton described the geology of the recovery site and the anatomy of the cranium, while Robert Thomson, the new professor of anatomy at the University of Cape Town (UCT), looked at the post-cranial

Figure 2.2: A photograph of Frederick William FitzSimons from his book on primates, published in 1911. He would have been about 40 years old at the time, but he appears younger in this photograph, so perhaps it was taken about the time he began work at the Port Elizabeth Museum in 1906. (Source: FitzSimons 1911)

remains, and the small fragments of archaeological material were described by Péringuey himself (Haughton, Thomson and Péringuey 1917). Smith's report on the brain followed directly after their paper, in the *Transactions of the Royal Society of South Africa* (Smith 1917). Based on his first assessment of the cranial vault shortly after the discovery, FitzSimons concluded that the skull was 'Mousterian and very closely allied to the original Neanderthal, if not the same race'.[10] The published report by Haughton, Thomson and Péringuey (1917) did not agree. Their decision was that although the remains did show some primitive characters, overall they were similar to the Cro-Magnon type. All the terminology used by Péringuey and FitzSimons was drawn directly from the European literature in the belief that the South African archaeology was reflecting the same evolutionary transition from primitive (Neanderthal anatomy and Mousterian technology) to advanced (Cro-Magnon anatomy and Aurignacian technology).

FREDERICK WILLIAM FITZSIMONS (1870–1951)

The key character in this small South African drama was F.W. FitzSimons, yet surprisingly little has been written about this man who was very much in the public eye, especially in the Eastern Cape Province, throughout the second and third decades of the twentieth century.

FitzSimons was born in Garvagh, Northern Ireland, but settled in Pietermaritzburg, Natal, with his parents when he was seven years old. Upon completion of his schooling, he returned to Ireland to study medicine, but after three years decided that a career in medicine was not for him. He returned to Pietermaritzburg in 1895 and worked first for the Natal Society Museum and then the Port Elizabeth Museum from 1906. It was in Port Elizabeth that he came into his own as the museum's director.[11]

FitzSimons's strength was public engagement. He was active in using the museum collections for educational purposes and in the two decades after he became director he sent out hundreds of small collections to schools to form the nuclei of school museums, even though this resulted in a serious depletion of the material kept at the museum (Plug 2020b).[12] Nancy Teitz of the East London Museum remembers seeing a newspaper article in the archives describing how FitzSimons arranged for a display of anatomical models 'demonstrated by young girls in white dresses at one of the Gala Evenings in the Feather Market Hall. They were an innovation at the time and almost as much of a sensation as FitzSimons' appearance on stage, swathed in snakes.'[13] Reptiles were FitzSimons's special interest and his Snake Park, launched in 1919, quickly became a major attraction in Port Elizabeth. He frequently made public appearances and wrote as frequently to the press and in popular magazines on scientific topics. Although he had no formal scientific training other than his abortive career in medicine, FitzSimons identified himself as an expert and was accepted by the public as such. His publications, other than the few that he sent to scientific journals and perhaps his books on snakes and snakebites, were generally at a popular level and have not survived as quality research in the South African literature. The idea that Groenewald would approach FitzSimons as an expert is not really surprising, given his public exposure, but in fact anthropology was not one of his subjects of knowledge.

Previously, FitzSimons had ventured into primatology and anthropology with his book *The Monkey Folk of South Africa* (FitzSimons 1911). The book was overly anthropomorphic and filled with unverified anecdotes, which are passed off as scientific observations. Although he did not go so far as to suggest that the Khoesan people were part of his 'monkey folk', his brief description of them comes perilously close. He already believed that 'the real true Bushmen ... are the descendants of the wild cavemen who came long ages and ages ago from Europe' (FitzSimons 1911, 97). In one of the Boskop letters, he tells Péringuey: 'I have myself studied craniology for a great number of years and am familiar with the crania of most present types of lower animals, and the various races of mankind past and present.'[14] In fact, before the finding of Boskop, FitzSimons had had absolutely no experience in archaeology and his anthropology was, at best, the result of a few physical anthropology lectures during his incomplete medical studies in Dublin.

The discovery of the Boskop skull opened up an entirely new field to FitzSimons. The brief experience of digging for more bones in Botha's trench was enough for him to begin to explore cave sites in the Tsitsikamma coastal region from 1915 onwards. Péringuey had sent James Drury, his own excavator at the museum, to the caves nearer to George in 1911, but Drury's careful excavation with detailed records and careful storage of excavated material did not look a bit like FitzSimons's work. FitzSimons's excavation techniques were extremely primitive, and it is still unclear exactly which sites were dug and how many skeletons were unearthed (Deacon 1979, 83). No clear records were kept about which caves he excavated and subsequent archival and field research has only located some of them (Schauder 1963; Turner 1970; Robinson 1977). Worse still is that we cannot even be assured which of the human skeletons came from which site (Morris 1992a, 5). The skeletons from FitzSimons's digs were nearly all transferred to the Department of Anatomy at the University of the Witwatersrand (Wits) in 1939 and although the catalogue system has kept a record of the original Port Elizabeth Museum numbers, it seems as if the transfer was disproportionately of cranial material. It is not clear if this occurred when the transfer was made to Johannesburg or whether FitzSimons did not keep most of the post-cranial skeletal elements he collected. Excavation records were not transferred with the skeletons and it is entirely possible that FitzSimons kept no records in the first place.

FitzSimons was mercurial and sometimes difficult, especially towards the end of his career. John Pringle was appointed as his deputy director in July 1935 by the board of trustees of the museum. When Pringle met with FitzSimons on his first day in his new post, FitzSimons said: 'I opposed the creation of this post because it is unnecessary. I don't want you here and there is no work for you.'[15] Pringle had been deputy for six weeks when FitzSimons quit. FitzSimons had built a large enclosure for live seals as part of the Museum's Snake Park, but the neighbours complained of the noise and he was forced to dispose of the seals.[16] FitzSimons was offended and abruptly resigned (Stuckenberg 2002, 51). FitzSimons was not well at this stage and

Pringle told Teitz that the museum board kept him on the payroll for two and a half years before finally officially retiring him and giving Pringle the post of director.

BOSKOP GOES PUBLIC

Péringuey, working with Haughton and Thomson, had been very slow and careful in his analysis. The conclusions showed none of the drama and showmanship of FitzSimons. Their 1917 paper simply showed that the European Cro-Magnons could also be found at the southern tip of Africa, but more importantly, that humans had a presence in the region deep in antiquity. But the understated conclusions were overwhelmed by Robert Broom, who submitted his own publication as soon as the first description came out. Broom (1918, 76) disagreed with Haughton's cranial capacity estimation, raising it to 1 960 cubic centimetres, a huge volume by any estimation.[17] Palaeoanthropology at the beginning of the twentieth century was convinced that it was the brain, especially its size, that defined humanity. Arthur Keith had proposed a minimum volume of 750 cubic centimetres as the marker of the genus *Homo* and Smith had suggested that a minimum of 1 000 cubic centimetres was required for the basic function of a modern human brain (Keith 1925, 603). Even though these researchers were aware that a large brain did not necessarily result in a clever man, they were convinced that the exceptionally large size estimates for the Boskop brain must have had a real meaning in terms of mental function.

Broom remarked that the cranial shape seen in longitudinal section matched several Neanderthal specimens and that the skull is 'at least as similar to Neanderthals as it is to Cro-Magnon' (Broom 1918, 78). The specimen could have been ancestral to both, making it as important locally as Piltdown was in Europe at the time. He emphasised this by giving the specimen a separate species name – *Homo capensis*. Broom, like most of the other scientists of the time, assumed that the

'Bushmen' had migrated down into South Africa and that the stone tools found by Péringuey were too early to be related to the San. In Broom's mind there was a very good chance Boskop people were the makers of the Stellenbosch hand-axes and therefore a representative of the earliest humans in South Africa.

In the meantime, FitzSimons had been busy disinterring human remains from cave sites along the south coast in the region of the Tsitsikamma forest. These were extremely rich cave sites with vast accumulations of food remains, stone tools, bedding and a large number of human burials. He published brief accounts of the excavations (FitzSimons 1921, 1923, 1926), but all of the published reports lacked significant detail. Of concern to FitzSimons was that there was a difference between the shallower layers and the deeper

Figure 2.3: Frederick FitzSimons excavating with the help of an unidentified labourer at Whitcher's Cave, near George, *circa* 1921. The photograph demonstrates how little archaeological context was preserved. (Source: Port Elizabeth Museum, with thanks to Emile Badenhorst)

layers (greater than 15 feet deep). FitzSimons interpreted this as representing two separate Strandloper populations: the more recent, higher in the deposits, whom he identified as 'Bushmen'; and an ancient more primitive group as 'Cliff-dwelling Strandloopers', who were earlier and deeper in the layers.[18] Ever the showman, FitzSimons used the newspapers to announce the excavations to the public.

In an interview with the *Cape Times* in October 1921, FitzSimons emphasised that although the caves had been inhabited by the 'Bushmen', they were antedated by his 'Cliff-dwelling Strandloopers'. He was certain that these people were Boskop and identical to the Palaeolithic people found in Europe. Their demise was due to the 'Bushmen' who penetrated the coastal regions, 'which up to that time had been exclusively the home of the cliff dwellers, and hunted the latter with their poisoned arrows, against which the primitive stone and bone weapons of the cliff men were impotent. Later came the Hottentots, who completed the work of extermination.'[19] FitzSimons presented his work to the papers as brand-new discoveries, but Péringuey was not impressed. In two subsequent newspaper interviews, Péringuey reminded readers that caves in this location had been explored before by Drury of the South African Museum and much, if not all, of the information had already been published by Péringuey in his 1911 monograph. He urged readers to look up Shrubsall's discussion of the skeletons, in which it was clearly evident that the prehistoric people were the ancestors, not competitors, of the ancient Strandlopers.[20] In an interview a week later, he recapped the story of the discovery of Boskop and gave a recount of Haughton's conclusions about the nature of the skull without giving it any special significance.[21]

FitzSimons forwarded some of the deep human remains to Raymond Dart in Johannesburg and it was he who made the formal connection between them and Boskop (Dart 1923, 625). Dart had been hired by Wits as its second professor of anatomy, arriving in

Johannesburg in January 1923. Dart's primary interest was in neuro-anatomy, a subject he had studied in London under Smith. He had in fact already seen the Boskop brain cast in Smith's laboratory when he was a demonstrator in anatomy at University College London (Dart and Craig 1959, 20). Although he was not keen to work on bones (Tobias 1984, 16), the arrival of the FitzSimons Boskop remains from the deep layers of the Tsitsikamma caves combined his latent interest in anthropology with his real interest in brains.

It says much about Dart's energy as a young researcher that he managed to get this first paper out in an entirely new field within his first year in Johannesburg despite all his teaching and administrative duties and the complexities of setting up a home in a new city. Dart carefully reconstructed the two fragmentary individuals from the deepest level of FitzSimons's cave sites and remarked on their similarity to the original Boskop skull. He especially remarked on the extreme length and breadth of the skulls and reconstructed a cranial capacity for the larger skull at 1 750 cubic centimetres, exceeding by far the average cranial capacity for European skulls (Dart 1923, 624). He commented on some Neanderthal-like and ape-like features although he was fully in agreement that the skulls were human. He disagreed with Broom's assessment of a new species, but he remarked that the skulls represented 'a race once widely distributed in South Africa from the Transvaal to the remotest south-eastern corner of the continent' (Dart 1923, 625). Dart began to subtly change FitzSimons's terminology, relegating 'Strandlooper' to the more superficial remains that FitzSimons had called 'Bushmen' and introducing the earlier group that FitzSimons had called 'Cliff-dwelling Strandloopers' as a separate Boskop race. He suggested that this Boskop race was separate from and preceded the Strandloper race historically. He then asked his newly appointed senior lecturer in anatomy, Gordon D. Laing from Aberdeen, and one of his promising students, Harry Sutherland Gear, to look at all the skeletons that FitzSimons was providing from his

Tsitsikamma sites. Laing looked at the more superficial 'Strandloopers' (Laing 1924, 1925), while Gear examined the deeper Boskopoid specimens (Gear 1925, 1926), and finally both of them wrote a summary (Laing and Gear 1929).

In the same manner as FitzSimons, Dart took to the newspapers to herald his discoveries. In an interview in *The Star* under the heading 'Older Than the Bushmen: Big-Brained Race, the Tzitzikama Discoveries – Popular Beliefs Shaken', Dart not only identified the Tsitsikamma remains as the same race as Boskop, but also linked them to the cave art. This he linked to the Cro-Magnon of Europe, acclaiming the South African peoples as highly civilised. Much of this echoed FitzSimons's preconceptions. We need to remember that Dart had been in South Africa less than ten months at this stage. Explaining how the 'Bushmen' had replaced them, he suggested in the same newspaper article that 'it might be they had discovered the poisoned arrow or some improved form of it, and so exterminated the intellectually superior but less well-armed race'. All of this was based on the assumption that the large brain case carried a brain that was functionally better than any other native group, putting these extinct people on a par with the ancient Europeans. Dart continued: 'It is disquieting to think that the gentle Bushman has been responsible for the extermination of a race that may, for aught we know to the contrary, have had a Dean Swift, a Shakespeare, a Rembrandt among its citizens.'[22]

Gear's work broadened the description of the Boskop type to include the post-cranial bones. He concluded that the bones from the deeper layers were 'neither Bushman, nor Negro, nor European' (Gear 1925, 468). The bones confirmed a robust body build. Gear then studied all six individuals from the deeper layers, confirming that all were similar to Boskop and that the Boskop race in South Africa was 'not a mere scattering of one or two individuals, but the existence of actual colonies of the type' (Gear 1926, 934). Laing had already compared his more superficial 'Bushmen' to the skeletons from the deeper layers and suggested that they had inherited some of the Boskop features by mixing with the earlier residents on their arrival (Laing 1924, 1925). He considered the features of Boskop to be an independent morphological type, which intermingled with 'Bushman' and 'Negro' types to produce the living aboriginal South Africans (Laing 1926, 908).

In the meantime, international approval for the Boskop type had come from William Pycraft at the British Museum and from Keith at the Royal College of Surgeons in London. In a long and detailed paper, Pycraft (1925, 193) compared the Boskop remains to early Europeans. He argued that it would be incorrect to call Boskop either Neanderthal or Cro-Magnon, but that it was (in opposition to Broom) a divergent branch of the species *Homo sapiens*. Although the pre-frontal area of the brain of Boskop was larger than that of a Neanderthal, it was smaller than that of the Cro-Magnon race. This last point is critical because at the time it matched the belief that the size of brain structures was directly linked to brain function and it implied that this large-brained African was not as intelligent as his European cousins. Keith felt the Boskop skull was important enough to devote twelve pages to it in his revision of his 1915 *The Antiquity of Man*. Like Pycraft, he did not agree that it was a separate species of humanity. He concluded that 'the Boskop man should be regarded as an ancient member of the stock now represented in South Africa by Bushman and Hottentot' (Keith 1925, 367).

Saul Dubow (1995, 52) has noted the racial overtones in the discussion of Boskop. One of the themes that appeared at the time was the notion of the 'Bushman' as a degenerate form of human, who could not possibly have produced the wide range of beautiful rock art present in South Africa. This was consistent with the European belief at the time that modern Europeans were superior to other races. Instead, the art was attributed to Boskop because of their larger brain capacity.

To Dart, his students and a growing number of his colleagues, the Boskop type was a new race comparable to other living races of South Africa. Alexander Galloway became a particular convert, as did Lawrie Wells while he was still in Johannesburg. Matthew Drennan in Cape Town identified a modern cadaver from his dissection room as an individual who was almost a complete throwback to this Boskop ancestry (Drennan 1925, 432; Galloway 1937a, 38). The extension of an extinct race into the present was completed when Dart identified the presence of Boskop ancestry in a group of San from the northern borders of the Cape Province in the Kalahari Gemsbok Park (Dart 1937a, 180). Dart measured each individual, with special reference to cranial size and shape, and assessed each individual according to his or her percentage 'Bushman' or Boskop ancestry, calculated by their similarity to the Boskop cranial features. Features that fitted neither of these preconceptions were said to be signs of intermixture from Bantu-speaking, Mediterranean (Hamitic), 'Armenoid' or 'Mongoloid' peoples (Morris 2012, S157).

While Dart and Drennan were extending Boskop into the present, Galloway was tracing Boskop into the distant past. On the assumption that Boskop was associated with the Middle Stone Age, as defined by John Goodwin and Clarence van Riet Lowe (1929), Galloway's team had 'slowly been piecing together an imaginary picture of the skull of the Boskop physical type' (Galloway 1937a, 32). By definition, the Boskop physical type would be the ancestral form present in the Middle Stone Age, which gave rise to the people of the Later Stone Age and would therefore be ancestral to living Khoesan populations. Not only were FitzSimons's sites in Tsitsikamma included, but now so were the earlier specimens from the neighbouring cave site at Matjes River and the more mineralised, and therefore presumably more ancient, specimens from Fish Hoek Cave and Springbok Flats. All were included because they fitted the basic massively constructed cranial type exceeding 190 millimetres in length and having a cranial capacity above 1 500 cubic centimetres. In craniological terms, these skulls were dolichocephalic (long), chamaecranial (low) and pentagonoid (five sided when seen from above) in form. Galloway stated: 'We regard all these Middle Stone Age skulls as being normal variants of one physical type – the Boskop physical type' (1937a, 32). Boskop had now been transformed from an interesting and possibly ancient skeleton from the old Transvaal into a major racial line of humanity inhabiting South Africa in the distant past, but still contributing genetic lines into the present.

The final brick in the construction of the Boskop edifice was Galloway's identification of the Iron Age burials from Mapungubwe, in what is now Limpopo Province, as being direct descendants of the Boskop race (Galloway 1937c, 120). The Mapungubwe individuals were buried with a rich assortment of gold, iron and ivory grave goods on a hilltop site surrounded by stone walls. The significance of his argument should not be understated, as Galloway was saying that the wonderful Iron Age stone-walled sites, with their rich elaborate burials, could not be attributed to the ancestors of the living Bantu-speaking peoples of southern Africa (Hall and Morris 1983, 30; Dubow 1995, 101). Galloway also analysed the skeletal remains from the sister site of Bambandyanalo near Mapungubwe for his Doctor of Science thesis in 1937, deciding on morphological grounds that the crania from both sites were from people who pre-dated and were unrelated to the modern Bantu-speaking populations. He stated 'deliberately and with a full comprehension of its significance that there is not a single specifically negro feature in any skull hitherto recovered at Bambandyanalo' (Galloway 1959, 121). Perhaps Galloway did not perceive this as an explicitly racist argument, but in hindsight we can clearly see the denial of cultural achievement in its historic context. This was no better than the argument made that the San were degenerate people who had lost a more glorious past.

The Boskop race survived the 1930s and 1940s, but was becoming more difficult to define, as new archaeological specimens were recovered in the 1950s. The individuals lumped into the archaeological definition of Boskop were so diverse that it was nearly impossible to justify any unity that could be defined as a single type. Boskop became less of a population than a typological essence that could be identified by features rather than individuals. In 1931, Drennan had suggested that Khoesan adults, as a whole, retained childlike features. This would explain the relatively small face and strongly bossed cranial vault (which gave the skull a pentagonoid shape when seen from above), a point already noticed by Keith in his conversations with Shrubsall in 1907.[23] Drennan called this 'pedomorphosis', a term drawn from the work of the Dutch anatomist Lodewijk (Louis) Bolk (1929), who talked about the retention of juvenile anatomy in the adult as a mechanism of human evolution. Dart (1940, 17) expanded on this by defining two patterns: pygmaeo-pedomorphic – 'Bush people' who kept their childlike features in a small body; and giganto-pedomorphic – Boskop people who had a robust physique, but also had childlike features. Phillip Tobias (1955, 6) tried valiantly to untangle this knot of genetic strains among the people of southern Africa by including the giganto-pedomorphic Boskop as one of the hybridising lines, and tried again in his epilogue to the 1959 publication of Galloway's 1937 Doctor of Science thesis. Tobias's epilogue tried to summarise all the developments in physical anthropology between 1938 and 1958, but what he produced was confusing and difficult to understand because he himself had not yet grasped the changes that were already happening in the analytical methods of the subject (Morris 2012, S158).

What Tobias could not grasp was taken in both hands by his Cape Town colleague Ronald Singer (1958, 176), who looked at the range of variation in the individuals in the sample, rather than the presence of preconceived morphological types. In the case of Boskop, he showed that the Boskopoid skulls were simply a group of individuals at one edge of the total range of variation seen in Khoesan populations. Nearly all individuals accepted as Boskop were large, robust males and their special range of morphological variation could be comfortably explained in terms of Khoe, San and 'Negro' variation (Morris 1986, 3). What struck him was the general African nature of all of these skulls. He did not see the Khoesan as a separate race, but as one of the regional variants of African populations. Both Tobias and Singer were part of the wave of the 'new physical anthropology', which began in the 1950s. Central to this shift in interpretation was the focus on the process of change, rather than the classification of types (Morris 2005b, 135). The method of analysis also shifted from descriptive anatomy to complex statistical techniques. William (Bill) Howells had begun his large-scale sampling of craniological populations in the mid-1960s and he would go on to use complex statistics to compare human morphology on a global scale without using racial categories (Howells 1973). Phillip Rightmire produced the first of these new-style analyses for South African archaeological skeletons in 1970. In Rightmire's analysis, the Mapungubwe and Bambandyanalo crania fell comfortably into the range of Bantu-speaking peoples. What had become obvious was that the earlier typologists were pre-selecting skeletons to fit their preconceived assessment of types. The new studies demonstrated that the Boskop range of variation was only a small part of the total variation in Khoesan and South African Bantu-speaking populations. The Boskop type represented large individuals at the edge of variation and did not represent a distinct human population. We now know that there was a diminution of body build between 4 000 and 2 000 years ago (Sealy and Pfeiffer 2000, 652). It is certainly possible that large, more robust individuals were more common in the earlier coastal populations, but they represented the extreme of the range of variation, rather than the mode. The lesson of the new approach to the analysis of human remains is that populations do change in body build and sometimes in shape over time, but this is a phenomenon of micro-evolutionary processes and secular trends, not replacement by new types.

The entire Boskop episode had some lasting lessons that we can see in hindsight. One is the importance of amateur collectors and regional museums in the discovery and description of human fossils. Groenewald's decision to give the bones to FitzSimons in Port Elizabeth would bring the bones to the knowledge of science. This would be echoed just over a decade later by the chain of events involving a quarryman and manager at the Buxton Limeworks in Taung, north of Kimberley in the Northern Cape (Dart and Craig 1959, 3; Tobias 1984, 22). Their actions would enable Dart to describe the Taung Child and change the face of palaeoanthropology. Although not archaeologists themselves, the local museum directors and limeworks staff acted as middlemen in bringing the bones to the attention of professionals.

But the factor that transformed the bones from interesting finds to fossil celebrities was the role of newspapers. The *Eastern Province Herald* regularly reported on events arranged by FitzSimons at the Port Elizabeth Museum, but it was Dart who demonstrated real skill at using these publications to present his scientific opinions. *The Star* in Johannesburg related his speculations about the Boskop skull, but Dart actively used the newspaper to popularise his announcement about the Taung Child in 1925. He sent the editor of *The Star*, F.R. Paver, a prepublication copy of his paper to *Nature* and when the *Nature* publication was delayed, Paver ran his own front-page lead on the discovery on 3 February, a full four days before *Nature* presented the peer-reviewed article (Tobias 1984, 35). Parts of the *Star* article were republished in the *Rand Daily Mail* and the *Cape Times* on 4 and 5 February, respectively. The newspaper coverage provided instant notoriety for Dart – in South Africa and the world. The newspapers were again involved when ancient bones were recovered in early 1929 from roadworks on the Springbok Flats, about 130 kilometres north-east of Pretoria. The director of the Transvaal Museum in Pretoria asked Broom to study the remains and they released their findings in the Johannesburg and Pretoria papers in the first week of February 1929. The articles were quickly picked up by the *Cape Times* and the *Cape Argus*, with headlines speculating on the presence of a race of giants in the Transvaal 100 000 years ago.[24] Although still speculative, these initial reports were followed by a multi-page article in the *Illustrated London News* on 16 March 1929. Not only did the *Illustrated London News* provide overseas exposure, but its format also allowed substantial details to be published, including photographs. Drennan had already used the *Illustrated London News* to announce his dissection room Boskop skeleton in 1925, but he also chose this vehicle to provide rich detail about the discovery of the Saldanha skull by Keith Jolly and Ron Singer in 1953. The popular press provided an outlet for publicity in ways the formal scientific literature did not. Whereas the academic press frowned on speculation, the popular press welcomed it. The carefully considered and peer-reviewed papers in the scientific journals would build scientific reputations, but it was the newspapers and illustrated popular journals that would manage the way in which the public understood the importance of the discoveries.

BOSKOP IS STILL HAUNTING US

The story of Boskop should have ended with Singer's 1958 publication. By 1970, mention of Boskop was nearly entirely gone from the literature. The skull languished in the storeroom of the Port Elizabeth Museum and was separated from the post-cranial bones, which were subsequently lost. There was some discussion of trying to radiocarbon-date the specimen, but no action was taken. The Middle Stone Age was pushed further back in time as more precise dating techniques were developed (Jacobs et al. 2008, 733) and large Khoesan-like skulls from the early Holocene (as Boskop was now assumed to be) were seen as part of the dynamic local population change in the South African Later Stone Age (Stynder, Ackermann and Sealy 2007, 7). The poorly preserved and undated Boskop skull was now seen to be of too little importance for discussion. All the other more complete specimens put into the Boskop type by Dart, his students, Drennan and Galloway now fitted comfortably within the range of normal variation of prehistoric southern African peoples.

All this changed with the publication of a new book by Gary Lynch and Richard Granger in 2008, from which excerpts were published in *Discover* magazine in December 2009. The *Discover* piece was titled 'What Happened to Hominids Who Were Smarter Than Us?' (Lynch and Granger 2009). Lynch and Granger's book, *Big Brain*, was admittedly speculative and a summary of their combined ideas on human neuroevolution. Lynch and Granger are neuroscience experts and much of the book consists of information about the structure and function of the human brain, but to augment this, and to build an argument that modern humans are not the pinnacle of human brain evolution, they resurrected Boskop in its entirety. The kernel of their argument is that Boskop (or 'Boskops', as they call them) had a brain not only 30 per cent larger on average than modern humans, but also the pre-frontal association cortex was 53 per cent larger, giving these people a distinct advantage in intelligence. Lynch and Granger recognise that total brain volume in modern humans is unrelated to intelligence, but their argument is that differences in brain volume between evolving species in the human line are critical to expanded intelligence. This meant that Boskop would have to be a separate species.

Lynch and Granger use all the old literature without any consideration of its modern context and without any recognition of the criticism of Singer and more recent analysts. They note that the specimen had been ignored since 1958, but give no reason as to why, other than that it was an inconvenient truth that our ancestors were smarter than us. Although Broom was the only researcher to declare the discovery a new species, they present their arguments as if all the early researchers were in agreement with Broom. They accept without question that it was the Boskop people who were responsible for the rock art of southern Africa, despite the fact that any reading of South African archaeology firmly places the authorship of the rock paintings and engravings with San people. They repeat all of the old speculative ideas of Boskop being replaced by less advanced *Homo sapiens*.

Perhaps the weirdest element in the book is Lynch and Granger reporting on the presence in one of FitzSimons's digs of a 'carefully constructed tomb, built for a single occupant – perhaps the tomb of a leader or of a revered wise man [who ...] appeared unremark-able in every regard ... except for a giant skull' (Lynch and Granger 2009). Where this information came from is not given, but it sounds very much like one of FitzSimons's public talks, where truth was less important than public spectacle. No tombs have ever been excavated in South African coastal cave sites and certainly there is no record of a single skeleton being found in such a context. It is hard to under-stand how Lynch and Granger learned about this tomb, but there is a clue in the files of the Port Elizabeth Museum. In a series of mostly uncatalogued and unreferenced photographs is one picture of a large mound of stones with a cranium and several bones laid out in the form of a tomb (Figure 2.4). Examination of the photograph clearly shows that this is a post-excavation layout of various cap stones and grind stones, along with many smaller rounded stones. A particularly well-preserved cranium has been placed on top of these stones. Perhaps Lynch and Granger had access to this photograph, or one like it, and they made the assumption that this was the grave as excavated. Most of FitzSimons's photographs were staged for a photographer, with bones laid out in careful order on top of unexcavated substrate. This is the same technique he used in photographing taxidermy specimens to demonstrate animal species in his 1911 book on primates.

Figure 2.4: An unnamed cave in the Tsitsikamma region excavated by Frederick FitzSimons in the 1920s. The unearthed stones and bones were placed in a staged setting in order to make the discovery appear more dramatic. (Source: Port Elizabeth Museum, with thanks to Emile Badenhorst)

The impact of Lynch and Granger's work has been interesting. Some reviewers praised the book as a stimulus for debate, with the caveat that it will be open to criticism by experts (Kirby 2009, 196), but the popular electronic media were more accepting, especially after the excerpts were published online in *Discover* magazine. The pseudoscientific fringe took over and Boskops became the remains of aliens, or mixtures of humans and aliens, or lost hybrid 'Caucasoids' of a mysterious past (Martinez 2016).[25] Most of the online articles included pictures of the original Boskop skull, but then photographs were added of obvious hydrocephalus archaeological and modern individuals. A few even put some really spectacular science fiction reconstructions of what these aliens would look like.

Fortunately, serious scientists have not risen to the bait and Lynch and Granger's work has had no impact on the field. Indeed, among scientists familiar with the archaeology of southern Africa there has been no reference to it at all. Only John Hawks chose to counter it directly in the same popular format as Lynch and Granger. Drawing on his knowledge of the discoveries and subsequent research on human evolution in South Africa, Hawks presented a point-by-point demolition of this most recent interpretation (2008, 2010). Hopefully Hawks has nailed it shut and Boskop will remain part of the history of the science, rather than the focus of analysis.

NOTES

1   'Report on the Council Meeting of the Port Elizabeth Museum', *Eastern Province Herald*, 8 April 1914.

2   'Report on the Council Meeting of the Port Elizabeth Museum', *Eastern Province Herald*, 8 April 1914.

3   Iziko Museum, Cape Town, letter from Louis Péringuey to I.L. Drege, 15 April 1914.

4   Iziko Museum, Cape Town, letter from Frederick FitzSimons to Louis Péringuey, 19 June 1914.

5   Iziko Museum, Cape Town, 'Agreement between P.J. Botha Esq of Boskop, Tvl. and F.W. FitzSimons Esq of Port Elizabeth C.P. on behalf of the Port Elizabeth Museum', June 1914.

6   Iziko Museum, Cape Town, telegram from FitzSimons to Péringuey, June 1914.

7   Iziko Museum, Cape Town, letter from Sidney Haughton to Louis Péringuey, 3 October 1914.

8   Iziko Museum, Cape Town, letter from FitzSimons to Péringuey, 10 October 1914.

9   Iziko Museum, Cape Town, letter from Péringuey to FitzSimons, 20 October 1914.

10   Iziko Museum, Cape Town, letter from FitzSimons to Péringuey, 20 April 1914.

11   Port Elizabeth Museum, Gqeberha, Michael Raath, 'Introduction to Inaugural F.W. Fitzsimons Lecture', unpublished manuscript (1990, 1).

12   Raath, 'Introduction', 2.

13   Nancy Teitz, letter to author, 29 July 1991.

14   Iziko Museum, Cape Town, letter from FitzSimons to Péringuey, 20 April 1914.

15   Teitz, letter.

16   At this time the Port Elizabeth Museum was situated on Bird Street, one of the better areas of the city. It was later moved to its present location, near the beach at Humewood. Emile Badenhorst, letter to author, 7 January 2020.

17   Some confusion exists about this figure. Broom (1918, 78) says it is 1 960 cubic centimetres, corrected to 1 980 cubic centimetres, but later he changed it to

1 950 cubic centimetres (1923, 148). All estimates place the cranial volume well above the human male average, which is approximately 1 350 cubic centimetres.

18 To FitzSimons, any people living on the coastline and surviving on marine foods were Strandlopers. He thought of them both as a separate people in the sense of his 'Cliff-dwelling Strandloopers' and an economic group in the sense of his 'Bushman Strandloopers'.

19 'Pre-historic Inhabitants: The Predecessors of the Bushman, Cliff Dwellers & Their Ways – Mr. FitzSimons' Interesting Excavations', *Cape Times*, 17 October 1921.

20 'Origin of Cave Dwellers: Earlier Discoveries in the Tzitzikama, the Work of the South African Museum', *Cape Times*, 22 October 1921.

21 'Cave Dwellers of South Africa: Account of the Boskop Skull, Discoveries at the Coldstream Cave', *Cape Times*, 29 October 1921.

22 'Older Than the Bushmen: Big-Brained Race, the Tzitzikama Discoveries – Popular Beliefs Shaken', *The Star*, 8 September 1923.

23 Iziko Museum, Cape Town, letter from Frank Shrubsall to Louis Péringuey, 22 March 1907.

24 'Revealed by Fossils: Race of Giants in Transvaal, Remains 100,000 Years Old – Most Important Find Yet Made', *Cape Times*, 9 February 1929; 'Fossil Remains of a Huge Primitive Man: Lived during Time of Gigantic Buffalo, Said to Have Existed on Springbok Flats', *Cape Argus*, 9 February 1929.

25 'Boskops became the remains of aliens': https://longsworde.wordpress.com/ 2016/03/03/the-mystery-of-boskop-man/; https://steemit.com/history/@geton thetrain/boskop-man-ancient-humans-with-extraordinarily-large-skulls- how-smart-were-they; 'mixtures of humans and aliens': https://hubpages.com/ education/Boskop-Man-Why-Did-a-Species-of-Hominid-With-150-IQs-Go- Extinct.

3

Matthew Drennan and the Scottish Influence in Cape Town

The South African College (known as the University of Cape Town after 1918) launched its Medical School in 1911. The teaching of medicine in South Africa had been under active discussion for a decade before this and, in preparation, the University of the Cape of Good Hope (the regulating body that controlled the quality of high school education around the country) had negotiated with three Scottish universities to recognise the preliminary undergraduate courses in science offered in Cape Town (Louw 1969). When the first course in anatomy began in May 1911, not only was the course recognised as a Scottish equivalent, but the first professor, Robert Black Thomson, was himself a Scotsman trained in Edinburgh, as was Matthew Robertson Drennan, his demonstrator hired in 1913 (Slater 1999, 255).

Janine Correia, Quenton Wessels and Willie Vorster (2013) have looked closely at the relationship between the early South African anatomists and Edinburgh. In academic hindsight, it really does become obvious why the connection should be so strong. South African medical practitioners were largely trained in Edinburgh during the second half of the nineteenth century. Out of the fifty-odd Cape-born doctors licensed to practise medicine in the Cape Colony in 1882, half were Edinburgh-trained (Burrows 1958, 150). Not only was there a bias at the Cape for Edinburgh medical education, but there were also hundreds of Edinburgh-qualified doctors, Scots and non-Scots, available and looking for medical and academic practices around the British Empire. Why was this single medical school so disproportionately important in training doctors (and anatomists) in the late nineteenth and early twentieth centuries? To answer this question, we need to look a little closer at the medical school in Edinburgh.

EDINBURGH AT THE CUTTING EDGE OF ANATOMY

Academic studies of anatomy in Edinburgh arose out of the conflict between physicians and surgeon-apothecaries, which had been going on from at least the 1650s. Surgeons were not perceived as real doctors by physicians because their training was more practical whereas traditional medicine at the time depended on an education with the classical overtones of ancient Greece and Rome. Medical doctors were seen as men of cultured education, but surgeons were not. Both qualifications required the study of anatomy as a base. When the higher centres of education were still predominantly in the countries of southern Europe, the physicians held sway and the classical approach was the path taken by the schools of medicine as they developed during the Renaissance, but by the early eighteenth century the northern cities of Paris, Amsterdam, Leiden, London and Edinburgh had eclipsed the old south as centres of trade and finance. Along with the new wealth and power came the demand for the very practical training of surgeons for their armies and navies. Andrew Cunningham (2010, 121) has noted that Edinburgh was particularly well placed to take advantage of this change. After the union of Scotland and England in 1707, young Scotsmen could easily take advantage of moving to England for new career opportunities. One of the best career options was to enter service in the navy as a ship's surgeon, especially since after three years' service, the still relatively young officer could set up in practice wherever he might want to, irrespective of the regulations of local guilds and councils. This immediately set up an additional demand for training in anatomy in Scotland, where training was close to home and, as a result, cheaper. By early in the eighteenth century there was a range of private anatomy teachers giving courses in Edinburgh. The town council had already recognised these demonstrators of anatomy as professors in 1705, but the biggest change was the appointment of Alexander Monro at the university in 1720 (Turner 1933, 102).

Monro had been trained in both Paris and Leiden and was able to argue that his appointment created a formal anatomy school linked to the university rather than purely a commercial enterprise for training hopeful surgeons. Monro was an anatomist at Edinburgh until 1754 when his son replaced him. Eventually the chair passed on to the grandson, giving the Monro dynasty a total of 126 years in the department of anatomy (Turner 1933, 328). The three Monro generations, all called Alexander, are known to history as *primus*, *secundus* and *tertius*. We know that Monro *primus* was a teacher of excellence who helped to build Edinburgh's reputation as a school of anatomy and medicine, but his descendants appear to have been less contributory. Monro *tertius*, in particular, was apparently a poor lecturer who struggled to get students to attend his course instead of those presented by more popular extramural teachers, who were common in Edinburgh at the start of the nineteenth century (Comrie 1932, 495). Private anatomy schools had become very common in Edinburgh in the late eighteenth century and since they were on a commercial basis, attracting students was critical (Cunningham 2010, 121). Despite his lack of success in gaining students, Monro *tertius* was an active collector of crania and began an interest in anthropology in the department even before this branch of science had a name. In the end, the failing Monro dynasty actually increased the competition between the private schools and the formal school in the university, which was to lead to an event whose repercussions were to affect the teaching of anatomy for the next century. This was the Burke and Hare scandal of 1829.

The most popular of the extramural anatomy lecturers in competition with Monro *tertius* was Dr Robert Knox. The anatomy taught by Knox was not only better presented, but it was also more up to date and, at the peak of his popularity, Knox's class had 500 students (Comrie 1932, 500). The new anatomy had originated in post-revolutionary France under Georges Cuvier and his successors. The old anatomy focused on anatomical oddities and pathologies. It did not have a philosophical base. Cuvier and his academic rivals, Étienne Geoffroy Saint-Hilaire and Jean-Baptiste Lamarck, offered students a subject based in comparative anatomy that helped to explain why anatomical shapes appeared in the way they did. Their work utilised new methodologies, including histology, and their search for anatomical patterns across species gave a meaning to anatomy, which put it at the forefront of the development of science in the nineteenth century. This philosophical anatomy travelled to Edinburgh with the Scottish radicals and would ultimately shape the teaching of anatomy even in the capital city of London when the first medical schools opened there in the 1820s (Desmond 1989, 36).

There was something even more important that made Edinburgh a centre of medical education. The study of medicine in eighteenth- and early nineteenth-century England was expensive, religiously exclusive and poorly taught. The established medical schools in Oxford and Cambridge imparted the earlier ideal of the leisured and classically educated physician and restricted students to those who professed adherence to the Church of England. The Scottish schools welcomed non-conformist Protestants and Catholic Englishmen (Cunningham 2010, 121). Scottish universities were technically Presbyterian, but none of them insisted on religious tests for their students in the nineteenth century. As a result, between 1801 and 1850 Oxbridge graduated 273 medical students, but 8 000 were trained in Scotland during the same period (Desmond 1989, 34). The classes of Knox and his colleagues were full, but the number of available bodies for dissection did not match the high demand. An entrepreneurial spirit in the form of resurrectionists who stole the bodies of the recently dead from their graves to sell to the medical schools negatively affected the reputation of early nineteenth-century anatomists, but the last straw was when Knox was found to have been purchasing the bodies of people murdered by William Burke and William Hare precisely to provide bodies for the dissection halls (Townsend 2001).

Within three years the British government had introduced the Anatomy Act of 1832, which helped to regulate and supply the bodies of paupers to the medical schools (MacDonald 2010, 17). In the mid-1830s Edinburgh University made its own anatomy course compulsory, which ended private anatomical schools. Knox's reputation was utterly destroyed and he never dissected again after 1830. He left Edinburgh in 1842 and the Edinburgh College of Surgeons took away his teaching diploma five years later. He earned his living by producing journalistic and sensational publications, moving away from the idealism of his earlier years and ending with a vision of a racial determinism that interpreted all civilisation and revolution as being racial struggles (Desmond 1989, 37).

In the meantime, Edinburgh continued as a centre of medical education. In the 1830s Edinburgh trained about a third of all medical officers in the armed forces and the East India Company (Desmond 1989, 34). There was another major increase in student numbers in the second half of the nineteenth century, with 435 students attending the practical anatomy class in 1876–1877 (Horn 1967, 200) and an extra-academical school in the Surgeon's Hall under the auspices of the university began teaching students who were not formally registered in the university (Turner 1933, 147). Upon the retirement of Monro *tertius*, the chair in anatomy was given to John Goodsir, who began to rebuild the quality of anatomical education, but it was Goodsir's replacement by William Turner in 1867 that finally brought anatomy in Edinburgh back to the heights it had commanded in the 1820s.

There had been a significant change in approach to teaching staff in mid-nineteenth-century Edinburgh. During the eighteenth century, a professor needed only to be a good and active teacher to be considered a success. Many simply gave the lectures as required by their appointment and waited out their retirement, but by the middle of the nineteenth century professors were expected to be researchers as well as teachers: 'No professor who was not himself engaged in advancing the bounds of his subject could hope to be a satisfactory teacher' (Horn 1967, 200). Turner was a superb example of this new tradition. The department taught not only anatomical structure, but also took a special interest in evolution. Turner was asked to describe the anatomical specimens recovered by the Challenger Expedition in 1876. The voyage of the HMS *Challenger* was a four-year oceanographic exploration, supported in part by the University of Edinburgh, intended to explore the deep oceans, but the scientific crew also collected specimens of interest from the land ports visited during the voyage. Of importance to anatomy in Edinburgh were the human and marine mammal skeletal samples. The small collection of anatomical specimens accumulated by Monro *tertius* and Goodsir was growing and a proper museum was needed to house them. Turner was the chair of anatomy in 1880 when the Edinburgh Medical School moved to its new buildings in Teviot Row and he ensured that the new lodgings included a purpose-built anatomical museum. The new Anatomy Museum, opened with great ceremony in 1884, was three storeys tall, with the second and third storeys represented by wide balconies along the side of the great room. The roof had glass skylight panels, which lit the huge space of the central hall all the way from the roof to the ground floor. This was the heart of anatomy teaching and research where the students at Edinburgh trained. By the winter of 1885 there were 638 students attending practical anatomy in the department and making use of the new museum.

Turner was promoted to university principal in 1903 and Daniel John Cunningham succeeded him as chair of anatomy. Cunningham had trained in Edinburgh, but had spent most of his academic life at Trinity College in Dublin. Cunningham's research interests were strongly comparative and embryological, but the evidence of the increasing size of the human skeleton collection during his stay in Dublin suggests that he had a developed interest in anthropology as well. His major focus in Edinburgh was on practical dissection for medical students and the development of postgraduate instruction in the School of Anatomy. Cunningham died in 1909 and he was replaced by another home-grown Edinburgh graduate, Arthur Robinson. Robinson had a special interest in embryology, but he also increased the contents of the museum with casts of prehistoric human crania, simian and human skeletons and tracings and casts of comparative craniology (Turner 1933, 106).

The Edinburgh Anatomy Museum had (and still has) substantial craniological holdings, with at least 120 crania in the collection originating from sub-Saharan Africa. The catalogue entries indicate that it was the Edinburgh medical diaspora who were the gatherers of these specimens, but the Department of Anatomy did have specific funds that could be used for the collection and study of human skulls. William Ramsay Henderson left a substantial sum in his will of 1832 for 'the advancement and diffusion of the science of phrenology and the practical application thereof'.[1] The Henderson Trust paid for the publication of books on the subject, supported students and paid for courses of lectures. The Trust also paid for anatomy assistantships in, and visiting lectureships to, the Edinburgh Department of Anatomy, and many of the cranial specimens listed in the catalogue are noted as being donated by the Trust. Although the Henderson Trust still exists, it moved its focus to more general anatomy in the 1920s and supported research on neurology and the function of the brain after that.

This is the environment in which the students who would become the first anatomists in Cape Town trained. The classes were filled with young Scots- and Englishmen who saw Edinburgh as a medical school of high quality, and the teaching activities inculcated a philosophy of scientific research in the students that would be carried into their careers, both as medical practitioners and research scientists. Thomson, the first incumbent of the chair of anatomy in Cape Town, had completed his medical degree in 1905, having been taught at least perfunctorily by the great Professor Turner. This was followed by a research fellowship in anatomy between 1906 and 1908 under Cunningham and finally a series of annual postgraduate courses in anatomy under the supervision of Robinson (Tobias 1990, 331). Drennan was several years junior to Thomson, studying under the tutelage first of Cunningham, until his death in 1909, and then staying on as demonstrator of anatomy under Robinson from 1911. Robinson acted as one of his references for the Cape Town job in 1913. Turner was no longer actively teaching from 1903, but he would have been a frequent visitor to the museum while he worked on his pet research project about the craniology of the Scottish people (Turner 1933, 105) and it seems likely that both Thomson and Drennan would have known him during their respective years as anatomical demonstrators.

Thomson and Drennan were not the only products of Edinburgh trained by Turner, Cunningham and Robinson who filled anatomical posts in the United Kingdom and abroad. Edinburgh pedigrees occurred on the curricula vitae of the anatomy professors at Glasgow, St Andrews, Cardiff, Liverpool, Leeds, Birmingham and even Oxford,

Cambridge and the famous Middlesex Hospital in London. The list continues for teachers in the early years of the medical schools in Dublin, Belfast, Tokyo, Lahore, Sydney, Melbourne, Otago (New Zealand), Montreal, Toronto, Dalhousie (Nova Scotia), Winnipeg, San Francisco, Cairo and of course in Cape Town (Turner 1933, 103).

MATTHEW ROBERTSON DRENNAN 1885–1965

Thomson was hired by the committee setting up the Cape Town Medical School because he had demonstrated a research ability and was well liked by the South African students who studied under him

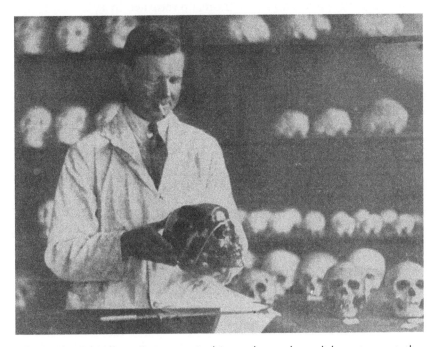

Figure 3.1: Matthew Drennan in his anthropology laboratory at the University of Cape Town in 1931. The cranium he is holding (UCT 74) is from the Later Stone Age deposits in a rock shelter from the Robberg site at Plettenberg Bay. The specimen was the topic of his article 'Pedomorphism in the Pre-Bushman Skull', published in 1931. (Source: *Cape Argus*, 27 August 1931)

in Edinburgh. He got off to a roaring start, introducing courses in anatomy, embryology and anthropology (Drennan 1963, 53). His short tenure as professor at the University of Cape Town (UCT) has been well described by Phillip Tobias (1990, 332), even including a speculation about the nature of the disorder that made him resign and take up a country clinical practice. The chief issue here is that for several months in late 1914 and early 1915, and again from November 1915 to June 1916, Thomson was on war duty, training medical officers and performing military clinical duties, and Drennan was called in as acting professor. Drennan's acting appointment was extended when Thomson's medical condition did not allow him to resume teaching all through 1917 and into 1918. Thomson returned in a part-time capacity, but formally resigned just after the start of the 1919 academic year. Although Thomson was responsible for setting up the anatomy teaching programme before the war, it was effectively Drennan who ran the course from 1915 to 1919, not Thomson.

The first six years of Drennan's appointment in Cape Town up to 1919 did not seem to have been particularly happy or fruitful and he quit, or came very close to quitting, on more than one occasion. He was very much the junior as demonstrator to Thomson, but the real problem was that the job itself seemed to be going nowhere. Although the courses that he was teaching were based on the solid curriculum and teaching experiences of Edinburgh, they were essentially service courses being taught in preparation for the second professional examination for students who would qualify in medicine elsewhere. There were delays in the parliamentary tabling of the University Act and when this was shelved indefinitely in 1914, delaying the launch of the university, Drennan resigned after only a year of teaching (Drennan 1963, 53). Private practice in Aliwal North occupied his time for the next year, but his time in the Northern Cape was not a loss, as he got to know the De Wet family, whose daughter Susannah he would marry in 1920. Early in 1915 Drennan volunteered for the

South African Medical Corps. His unit, the Brits Commando, was rushed to what was then South West Africa without preventative vaccinations, and as a result he contracted enteric fever (Drennan 1963, 53). He was invalided back to Cape Town, where he found employment picking up the absent Thomson's teaching load, and also doing some part-time surgery at Alexandra Hospital in Observatory (Louw 1969, 99).

As Thomson's absence stretched into 1917 and then into 1918, Drennan's professorship continued to be listed as 'acting' and he again made up his mind to leave. The teaching load had risen to over 300 students a year even though the courses had originally been planned to have a maximum of only 50. Drennan complained that he serviced these students even though he was given 'no qualified assistance' (Drennan 1963, 53). UCT had finally been established by an Act of Parliament in April 1918 and the first clinical courses were scheduled to be offered in 1920, with the first class to graduate at the end of 1922, but the Faculty of Medicine had still not decided about Thomson. When Drennan heard that a medical school was being launched in Johannesburg, he decided to apply for the professorship there. At almost the same time that Thomson submitted his formal resignation, Drennan made his intention to apply for the Johannesburg post known to the General Purposes Committee and asked for their support in his application. The board of the Faculty of Medicine set up a team to fill Thomson's chair and in just two days recommended to the council that Drennan be appointed professor of anatomy without any advertisement (Tobias 1990, 334). As Tobias points out, their 'remarkable agility' was impressive. Drennan agreed to stay on.

What is interesting about Drennan is that during this early period he expressed no interest in research, yet we know from his later career that research was an important aspect of his activities. When and where did Drennan develop his interest in research, especially his anthropological research?

Drennan grew up in Galston, Ayrshire, south of Glasgow in Scotland. He was an excellent student who won the George Heriot Classical Bursary to Edinburgh in an open competition. Drennan's university career was impressive, earning him several bursaries and honours (Drennan 1913, 3; Louw 1969, 99). The Heriot Bursary enabled him to complete a Master's degree in classical and modern subjects in 1907 and then to proceed to medicine, which he completed in 1910. He worked as a junior demonstrator of anatomy under Robinson and was entered into the Fellowship of the Royal College of Surgeons (Edinburgh) in 1912, with advanced anatomy and embryology as speciality subjects. Although he was a stellar student and excellent lecturer, we have almost no evidence of his interests in research at this time, with one exception. During 1912, Drennan presented a paper to the Anatomical Society of Edinburgh on 'the homologies of the limbs' (Obituary 1966, 122). The paper was not published, but as we shall see, Drennan revisited the topic some six years later when we begin to see the first real evidence of an interest in research.

Although the years 1918 and 1919 were overwhelmed by his teaching commitments, there are some shards of evidence that suggest Drennan was beginning to think about pursuing research. The anatomy course at the South African College had been set up as a two-year programme (Anatomy I and Anatomy II), leading to an opportunity to write the professional examinations at other schools. Students taking this option would receive credit for their Cape Town work, but no degree. Once the clinical subjects had been planned for launch in 1920, students could also register for a three-year Medical Bachelor of Arts, which would qualify them for entry into the clinical years, but also give them a self-standing degree in anatomy (or physiology), with an option to continue for a further year to earn a research Master of Arts in the subject. The anatomy courses were tough and roughly one-third of those who attempted Anatomy I were unsuccessful. Of the remainder, only about a quarter chose the option of the separate Bachelor of Arts, with the opportunity to complete a Master's if they passed Anatomy II (Louw 1969, 100). The bulk of students who chose the research option went into physiology. This reflects the well-documented evidence that William Jolly, the professor of physiology, was actively developing research while Drennan in anatomy was not (Louw 1969, 101). The first of the Medical Bachelor of Arts' students to attempt the one-year research Master's in anatomy was Vincent Vermooten in 1919. He chose an anthropological topic – 'the long bones of the South African bushmen'.[2]

Was it Drennan who suggested the topic to Vermooten? The latter provides no information on this, but it may very well be that it was Vermooten who brought anthropological research to the attention of Drennan, rather than the other way around. The peak of 372 in post-war student numbers was reached in 1919, and Vermooten was hired as a student demonstrator to help Drennan deal with this almost impossible teaching task. He worked on his project while helping in the dissection hall and his primary contact was Dr Louis Péringuey at the South African Museum. Vermooten accumulated data on 38 Khoesan individuals, all but five from the South African Museum and the balance from the Department of Anatomy's skeletons, collected from the Richtersveld by Winnifred Tucker a few years before (Morris 2000, 74). In a letter to Frank Shrubsall in January 1921, Péringuey complained that although the Medical School was now up and running, Thomson had left academia and there was no one to continue the studies.[3] At this point Drennan had been professor for two years and yet Péringuey did not even mention him. Péringuey told Shrubsall about Vermooten in a subsequent letter, although he did not mention him by name.[4] He said that Vermooten had gone to Johns Hopkins University and Péringuey had just given him permission to publish the data he gathered at the museum. Never does Péringuey mention Drennan as a possible substitute researcher and it is clearly evident that, at least in the summer of 1921, Péringuey did not consider Drennan to be an academic with an interest in anthropology.

Vermooten had indeed completed his studies in Cape Town and had moved on to study medicine at Johns Hopkins University in Baltimore. He completed his specialisation in urology at Johns Hopkins and had a long career as a clinician in the United States. He published several papers on urological topics, but only one on an anthropologically related field. His anthropology paper (Vermooten 1921) was a description of the damage to the cervical vertebrae of four individuals who had died by hanging.[5] The reason that his papers have survived is that in 1921 he wrote to Drennan, sending him his old rough notes and thanking him for his comments on the long bone paper, which was still under revision.[6] Drennan was clearly closely involved in helping Vermooten to publish his work, but he was not particularly familiar with the content. Vermooten said that it was his plan to submit it to the *American Journal of Physical Anthropology*, but we have no record of it being submitted.[7] Drennan did send a copy of the unpublished final draft to Eugène Pittard in Geneva. In his letter to Pittard, Drennan mentioned that the paper had been submitted to the *Transactions of the Royal Society of South Africa* (presumably in 1919 or 1920), but the reviewer, Robert Broom, had rejected it because he did not agree with the classification of Bushman peoples used.[8] Shrubsall and Péringuey had argued that the Khoe and San were closely related, but the coastal Strandloper groups were separate populations. Broom, on the other hand, argued that the Khoe and San were visibly different and that Strandlopers were just Khoe or San who were living at the coast. It did not help that Péringuey and Broom were not on speaking terms (Findlay 1972, 27). Vermooten was using Péringuey's classification system, so was seen by Broom as being in the Péringuey 'camp'.

One of the rough drafts that Vermooten sent to Drennan in 1921 reveals a bit more about Vermooten's research, but much more about the potential for anthropological research at the Department of Anatomy at UCT and how Drennan was drawn into it. In August 1919, while Vermooten was a student demonstrator, the cadaver of a 25-year-old woman was brought into the department for dissection.[9] Based purely on the appearance of her physical traits, the department identified this young woman as 'Bushman' or 'Hottentot' and she became a specimen of specific interest. Vermooten took extensive notes and measurements and it was his intention to write up his description as a paper.[10] The description talks about 'simian characters' and 'purity of breed', and has a special focus on regenerative organs, so it was obvious that the young woman was not being seen as part of the range of variation of the peoples of the Cape, but instead as an example of racial type, with all of the negative attributes assumed by racist perceptions of Khoesan peoples at the time. The fact that race was something being examined when cadavers were accessioned for dissection became obvious when in 1925 Drennan announced the finding of the corpse of a contemporary Boskop man in his dissection hall. This was Drennan's first big anthropological contribution to research and it clearly implies that racial origins were a focus of interest. The skeleton of the woman from 1919 was not kept as a research skeleton because Drennan had not yet begun to think in terms of an osteological research collection.[11] But not so for the Boskop man in 1925.[12] By the middle of the 1920s Drennan had developed a thriving interest in the anthropology of South African peoples and had begun his long career in the anthropological tradition of Turner in Edinburgh, but there was little sign of it happening before the Boskop publication.

An obvious question is whether this anthropological interest was entirely new or whether it was predetermined by his Edinburgh training. There is one clue that suggests Drennan was expressing his Edinburgh experience of a broad definition of anatomy in which anthropology was a subject of interest, but not a focus of research. Drennan had special display cabinets produced to house his growing collection of anatomical specimens. These beautiful wooden and glass cases are still present in the department at UCT. They stand two metres tall and are a copy of those found in Turner's Edinburgh Anatomy Museum. Their contents reflect not a systematic anthropological focus, but an eclectic mix of anthropological and anatomical specimens, including items of pathological interest. There is no overriding theme to the early anatomical specimens accessioned in the 1920s and, as in Edinburgh, the collection is more of specimens obtained as opportunities arose. Most were donations, but there were also traded specimens from other anatomists around the world. Drennan corresponded with J.C.B. Grant, then at the University of Manitoba in Canada, in order to obtain a good example of a Canadian 'Indian'. Grant provided Drennan with a complete skeleton of a First Nation Canadian from the dissection hall in Winnipeg, and Drennan reciprocated with several skeletons of 'Bushmen', a trade that was reversed in 2012 when the skeletons were returned to their respective countries of origin.[13]

The embryological material is a good example of how specimens were accumulated. Drennan kept a file of information on specimens of developmental interest, including many of the original letters that accompanied specimens.[14] Between 1918 and 1935, Drennan added 21 foetal specimens to his display case, all of them donated by practising physicians. That embryology was an interest of Drennan's is incontestable. Among his very first publications was a brief note to the *South African Medical Record* (Drennan 1922a) on the scope and aims of embryology and in the same year he published a set of course notes on human embryology for medical students (Drennan 1922b). He followed this with a privately published series of embryological illustrations (Drennan 1927c). The *Short Course on Human Embryology* went to a second edition in 1930, a third edition in 1939 and finally a fourth edition in 1950. But there are no research papers published on the topic by Drennan or his students. The only paper of note is an unpublished descriptive essay on early human development by Sam Scher, his Master's student in 1926.[15] The specimens themselves were mostly of normal development, but from ectopic pregnancies or stillbirths. The accompanying correspondence clearly shows that the donors felt that donation of the little specimens to the medical school was in preference to their disposal as medical waste.[16] We have no idea how the mothers felt about the donations, or if they even knew this was happening. Whatever the motive for donation, it is evident that Drennan did not consider them of research interest, but only as examples for the teaching of embryology and something of value to store in the anatomical museum.

Although Drennan had not yet begun his research in anthropology at the time Vermooten was exploring his racial studies and the embryological specimens were starting to come in, one letter in the UCT files suggests that he was beginning to think about research and research publication. Sometime in the winter of 1918 Drennan had pulled out his notes on the homology of the limbs and was putting them into order for publication. He sent a draft to his old adviser in Edinburgh, and on 1 August Robinson responded with a long letter full of advice. Drennan was proposing a new theory he called looking-glass similarity or mirror image homology. For those unfamiliar with this subject, we have known since the nineteenth century that humans, like all vertebrates, are segmented from head to tail in the same manner as earthworms, but in worms the segments are broadly undifferentiated, whereas in humans the segments have been altered so much over evolution that it is difficult to see the underlying pattern. The arms and legs each arise from one segment, but then develop what has been called serial homology – showing a series of parallel structures as we look further along the limb from the central body. The question has been whether or not the forelimb and hindlimb follow the same sequence. Drennan agreed that they do, but he argued that the arm and leg are not simply parallel segments, but mirror images of the segments.

From his days in Edinburgh, Drennan had developed an interest in deep fundamental questions of evolution, as seen in anatomy. The draft he sent to Robinson was intriguing enough for Robinson to respond in detail, but he thought that Drennan was 'trying to make your facts carry a greater burden than they are capable of entertaining'.[17] Robinson told Drennan that his paper was not publishable as it was, but a more general paper might provoke some interesting discussion. Drennan still needed to learn how to put his arguments into a logical scientific format, but this correspondence shows that Drennan was now serious about research. In the end he did indeed publish his paper (Drennan 1927b), but only after he had begun to focus on more specific research interests, including anthropology.

DRENNAN'S ANTHROPOLOGICAL LEGACY

It took a decade for Drennan to become a productive researcher. By 1925, at the age of 40, he had begun to produce research papers regularly. There were three broad areas that interested him: clinical medicine, evolutionary theory in anatomy, and anthropology.

Drennan's first publication was a brief note in 1921 to the *South African Medical Record* on the need for clinical research in South Africa. This was followed by a further eight articles in the next three years, which were either brief notes on clinical or anatomical topics or short renditions of lecture notes for students. None were data-based explorations. Then in 1925 and 1926 there were two papers on anthropology: a brief announcement in the *Illustrated London News* on the finding of a Boskop skull in the dissecting room at UCT (Drennan 1925) and a paper on the anatomy of Khoesan people (Drury and Drennan 1926) based on observed data. It is evident that Drennan had become focused on anthropology, but who was influencing him?

The first person of influence was undoubtedly Raymond Dart. Both Dart and Drennan were called upon to act as co-examiners for the annual second professional exam in anatomy at their respective universities. Drennan jokingly called these annual meetings the South African Anatomical Association because they were the only professors of human anatomy until the University of Pretoria began teaching medicine in 1944 (Dart 1960, 828). Dart called Drennan his academic companion and elder brother. Dart had begun anthropological research upon his arrival in South Africa in 1923 and anthropology was a topic the two of them must have discussed socially between bouts of student examining. In a private letter to Sir Arthur Keith in 1929 about anthropological research, Drennan referred to researchers who came up with grand theories as 'live-wires' and then added 'certainly true of my colleague Dart'.[18] Dart's excitement about the creation of the Boskop type must have been behind Drennan's identification of a contemporary Boskop on his dissection slab in 1925.

The second influence was from his science students. Vermooten brought Drennan into contact with the Khoesan skeletons at the South African Museum and although Vermooten was unable to publish his work, Drennan became his spokesperson in relations with other scientists, sending the original or copies of Vermooten's project not only to Pittard in Geneva, but also to Rudolf Martin in Munich and to Dorothea Bleek in Cape Town. Bleek, one of the mothers of South African ethnology, wrote back to Drennan saying that she found Vermooten's work very interesting, but she disagreed with the classification of 'Bushmen' in the paper. She wrote:

> I understand that the grouping of the Bushmen into Strandloopers and Inland Bushmen was done by Shrubsall, and seems to have been followed more or less by the staff of the S.A. Museum. Of course I am ignorant of their reasons for such a division, but it seems to me a very arbitrary one. It is as if in classifying the inhabitants of Great Britain, the inhabitants of Kent and Sussex were put into one group, and all of the rest up to the Highlands of Scotland into another.[19]

This letter from Bleek arrived at just about the time that Drennan was beginning a collaboration on Khoesan people with the museum taxidermist, James Drury. Under Péringuey's direction, Drury had been travelling widely in the northern Cape and South West Africa, making casts of disappearing aboriginal people. Between 1909 and 1921 he made moulds of 68 people (nearly all of whom were Khoe or San) and kept detailed records of their bodily characteristics, especially their genitalia and skin colour (Plug 2020c). On his last two trips in 1920 and 1921 he was accompanied by Dorothea Bleek. With Péringuey's death in 1924, Drury had no one to work with at the museum who had physical anthropology experience and it is obvious that Drennan was stepping in as adviser in the same way he was helping his students. This initial work with Drury was a matter of helping him make sense of the anatomical structures (particularly sexual) that he had noted when making the casts, but the partnership would grow into osteological analysis when Drury began his dig in the 'Bushman cemetery' at Colesberg in 1926.

Vermooten was just the first of a series of science students under Drennan's supervision. The year after Vermooten graduated, David Lurie completed his Master's in anatomy with a set of detailed dissections of the Cape baboon, followed the next year by Max Gorfinkel working on the cerebral hemispheres of the brain and then in 1922 with Otto Hooper completing a project on the structure of the heart. Gorfinkel so enjoyed his course that he stayed on as anatomy lecturer in 1922. We have no record of students during the next two years, but 1925 was a bumper year, with no less than three students completing the Master's in anatomical science. Solly Cohen did a project on the brain of the Piltdown man, using the shape of sulci on the inside of the occipital bone to recreate the original soft tissue structure. Drennan actually wrote this up for publication (1927a), giving Cohen credit for the original data gathering. Thomas Dry worked on the calculation of stature from anthropometrics and Solly Zuckerman produced such a superb project on the cranium of the baboon that it was published with minimal editing in 1926. Cohen came back into the department as a part-time lecturer between 1933 and 1938. Dry moved on to study medicine and eventually worked at the Mayo Clinic. Zuckerman became an anatomist in the United Kingdom, later earning a knighthood for his administration of research.[20] The mix of anatomy and anthropology in student projects continued in the late 1920s and 1930s, all of which had Drennan as a supervisor, with several leading to publications (Slome 1927, 1929; Slome 1932; Glaser 1934). These latter projects coincided with the most productive period of Drennan's research and also the growth of the research collection of anthropological and anatomical specimens in the department.[21]

Drennan produced 103 publications during his lifetime as an academic and it is relatively easy to follow his line of anthropological inquiries from his list of technical and popular articles. After the first brief foray into anthropology with Drury's casts and the Boskop type, he then dabbled with the brain of the Piltdown discovery from cranial endocasts sent from London and more seriously with analysing the skeletons from Drury's new dig at Colesberg. Drury had already had extensive experience of excavating on the Cape south coast, but in late 1926 the South African Museum was informed of a large series of graves outside of the town of Colesberg. Local informants indicated that they were the graves of 'Bushmen' who had died in the great smallpox epidemic in 1866 and Drury excavated the graves as part of the museum's programme of accessioning San skeletons.[22] Drury asked Drennan for help in the analysis of the bones. Drennan assigned the examination of the skeletons to one of his students, David Slome, but took on the job of looking at the teeth himself. The work was done in 1928 and resulted in two publications in the *Annals of the South African Museum* (Slome 1929; Drennan 1929b). Drury's original assessment of the site (sent as a note to Slome and included verbatim in the 1929 paper) indicated that the skeletons were actually from a mixed community, rather than a pure San group, and Slome's work in particular confirmed this.[23] Drennan selected the individuals who were more Khoesan in morphology and made formal reference to them as a 'Bushman Tribe'.[24] This was consistent with Drennan's developing interest in racial typology and most specifically in his interest in Khoesan peoples in particular.

As Drennan's work on the skeletal remains became more widely known, his laboratory at the UCT Medical School became the destination of choice for analysis of new discoveries, mostly from the Cape Province, but also for discoveries in Rhodesia (Obituary 1966, 123). It was while the writing up of the Colesberg project was under way that Drennan got his first chance to be the primary excavator of an ancient specimen. News reached him of old bones found in the bottom of a sand quarry on the Cape Flats just outside of Cape Town. Sadly, all of the original details of the discovery have been lost. Drennan was meticulous in keeping correspondence and site records in file boxes for each of his projects, but in 1954 and 1955, Ronald Singer, then a senior lecturer in the Department of Anatomy, borrowed seven boxes and never returned them.[25] The result is that the history of the discovery must be reconstructed from the few surviving snippets of correspondence and the limited details in Drennan's published works.

The Cape Flats discovery was Drennan's first ticket into the international theatre of fossil hominid discoveries. The circumstances of the finding are outlined in Drennan's paper for the *Journal of the Royal Anthropological Institute* (1929a), although there are many details missing. Drennan was taken to the quarry by Gerald McManus, a technical assistant in the Department of Physiology from 1920 to 1940 (Louw 1969, 116), sometime earlier in 1929 and what they discovered was an active quarry a little over four metres deep at its deepest point. The workmen in the quarry had set aside various bones, stones, artefacts and ferrocrete nodules, which they had found while sifting sand to purify it for building purposes. Drennan found other pieces of bone, teeth and implements scattered on the floor of the quarry. He was able to reconstruct at least one large partial skull. Subsequently, he and several volunteers sieved the loose sand below the discovery and further fragments of at least two or more individuals were recovered, including a frontal and temporal fragment that fitted his reconstructed skull. What excited Drennan was that his new skull was mineralised, suggesting great antiquity, and very robust, with strong

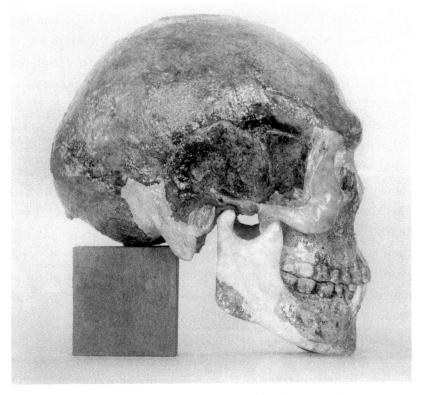

Figure 3.2: The Cape Flats skull, discovered by Matthew Drennan near Cape Town in 1929. This was the first fossilised human specimen described by Drennan and helped to build his reputation as an anthropologist, rather than an anatomist. (Source: Department of Human Biology, University of Cape Town)

brow ridges, large teeth, and 'hardly any Bushman characters what-soever' (Drennan 1929a, 423). In his opinion it showed Australoid features and confirmed Broom's idea of an ancestral line of robust Neanderthaloid individuals in South African prehistory, similar to Australian aboriginals (Broom 1923, 146; 1929, 507). The fact did not bother Drennan at all that the finds included skeletal elements similar to modern Khoesan and a range of artefacts from Middle Stone Age to modern historic. In the same manner as he dealt with the range of variation at Colesberg, he simply ignored anything that disagreed with an ancient origin or suggested Khoesan anatomy.[26] He sought guidance from Keith about publication of his new and important find.

Drennan had begun a long correspondence with Keith in 1926.[27] He had sent Keith a copy of his paper on the anatomy of the human heart and would continue to send Keith copies of his publications over the years. This relationship connected Drennan to the world of science outside of the small community in South Africa, and in return Keith gained a correspondent who was intimately familiar with recent anthropological finds in South Africa. As far as the correspondence goes, it appears that the two men never met, and Drennan was extremely disappointed when Keith was forced to cancel his visit to South Africa in 1929 due to unforeseen circumstances.[28] Shortly thereafter Drennan wrote to Keith, sending him a copy of his draft paper on the Cape Flats skull, asking if Keith thought it might be appropriate for the *Journal of the Royal Anthropological Institute*. Drennan did not want to publish it locally because 'there would be no one to criticise it, and few to read it'.[29] In addition, Broom had published his 1923 paper in the journal and Drennan's work would be a sequel, adding currency and criticism to the idea of an Australoid race in South Africa. Keith agreed and forwarded the paper to the editor of the journal with a recommendation for publication. Keith was intrigued by the specimen and gladly accepted Drennan's offer to send him the skull for observation. Drennan arranged for the skull to be hand-delivered to Keith by his student Isidore Zieve, who was travelling to London to continue his medical studies.[30] Zieve not only acted as the delivery person, but was also able to give Keith a guide to the map location and local setting of the Cape Flats and Fish Hoek discoveries (see chapter 2).[31]

In Drennan's opinion, the Cape Flats skull was the link that connected ancestral humans in Europe to their African contemporaries. Writing in his textbook on physical anthropology, Drennan says: 'It presents a good example of a type, which is somewhat intermediate between Mousterian and modern man' (Drennan 1930a, 46). He goes on to link the recently discovered Fish Hoek skull and various Strandloper specimens as being part of a local variant of the European Upper Palaeolithic Cro-Magnon race. Drennan's Cape Flats skull, despite the almost complete lack of association with any archaeological context, became a major plank in the typological assessment of South African fossil specimens into robust Neanderthal-like types and gracile Bushman-like types. Drennan went on to link these types into the developmental mechanism suggested by Lodewijk (Louis) Bolk (1929) that was being used to describe the characteristics of the Boskop skull. Drennan took Bolk's terminology and divided prehistoric South Africans into paedomorphs (child-like) and gerontomorphs (adult-like). These terms would be entrenched in South African anthropology until the 1960s.

Drennan's anthropological papers between 1929 and 1931 would cement his international reputation as an expert on the origin of Khoesan peoples in the region and would trigger him to write an even wider range of anthropological papers in the following years. His archaeology was self-taught and mostly from a Eurocentric perspective, often incorrectly trying to shoehorn local cases into an explanation drawn from the international literature (Drennan 1930b, 547; 1937, 183). When Donald Bain brought a group of southern Kalahari San to Cape Town as part of the United Empire Exhibition of 1937,

Drennan was asked to help in their scientific description. One woman in particular, /Khanako, was cast under Drennan's direction, and her body cast, along with the plaster copies of her head, foot, hand and genitalia, were added to Drennan's museum collection (Rassool and Hayes 2002, 127). He continued to provide reports on skeletons from archaeological contexts, including the well-excavated and rich site of Oakhurst dug by the UCT archaeologist John Goodwin (Drennan 1938a, 1938b). Drennan and Goodwin worked closely together on the analysis of the skeletons, but Drennan had no role in the archaeology.

DRENNAN'S COLLEAGUES

Although the 1930s were very productive for Drennan in terms of anthropological and anatomical publications, the number of students studying anthropology during his tenure was less impressive. Drennan tended to work alone and he preferred teaching anatomy to anthropology. This should really not be a surprise as his training was in medical anatomy and his anthropological work was overshadowed by that of Dart. The few students that did stay on to do the Master's in anatomy under him invariably left to pursue a medical career, and the failure of the university to provide an additional senior permanent post in anatomy meant that Drennan shared the burden of teaching with temporary lecturers, who stayed for too short a time to have an impact. Drennan lamented this in a letter to his former student Jack Heselson, who had helped to teach anatomy in 1935 and 1936, but had since departed to complete his surgical training in London:

> Since you left, things have not gone too well amongst my assistants. Dr Shapiro's enthusiasm for anatomy evaporated very quickly, and he left with rather short notice at the beginning of this year. He developed some gastric ulcer or other tummy complaint but this was complicated by falling in love or some other equally demoralising

trouble. Dr Downes has stepped up and done splendidly so far, but her husband has been transferred to Mafeking and she leaves at the end of September. Mr Tucker BSc, one of my science men and a Rhodes Scholar, has acted as a very good standby in the junior post, but he leaves shortly for Oxford. I understand a posh lectureship is being advertised for next year, but meanwhile I am the sole survivor amongst the full-time staff.[32]

The 'posh lectureship' did indeed materialise and in 1939 the department hired Dr J.A. (Jack) Keen as senior lecturer in anatomy.

Keen (1894–1969) was a skilled ear, nose and throat surgeon. He had been born in Hong Kong to a Swiss/German missionary doctor and trained in London at Charing Cross Hospital. Keen joined the British Medical Corps in the First World War and was sent to East Africa where he contracted malaria. He was sent to Cape Town to convalesce and he so much enjoyed his time there that he made up his mind to find employment in Cape Town if it became available. Keen took up private practice in Leicester, but he did not enjoy the clinical experience and it was around this time that he began to think about anatomy as a profession.[33] When the opportunity arose at UCT in 1939, he accepted it and relocated his whole family (excepting his son, E.N. [Ted] Keen, who was following his father's footsteps and studying medicine in London). Keen was very different from the part-time and temporary lecturers that Drennan had shared his department with until then. Although all of his research was clinically based, Keen had published nearly twenty papers in technical journals by the time he arrived in South Africa. The appointment in Cape Town perfectly suited Keen and he began almost immediately to make use of Drennan's eclectic museum collection. He brought with him a specific interest in embryology, teratology and congenital malformations, but the collection of anthropological material in Cape Town really sparked his interest. Within a year, Keen was publishing more papers than Drennan, his superior in the chair of anatomy.

Keen was focused and determined as a researcher. Singer, who studied under him as an undergraduate, saw this as aggressiveness, but not in an unpleasant way. Singer recounts that when Keen had decided to go to Brazil on a research trip, he spent a year in advance studying Portuguese: 'When I say study, I mean every minute of the day where he wasn't doing work, teaching or doing his research work, he had a book on Portuguese somewhere near him, he had Portuguese comics, he had Portuguese Reader's Digest, he listened to Radio Mozambique.'[34] Ted Keen remembered his father as a workaholic who gave his all to any project he undertook.[35] The skeletons in the department (both archaeological and cadaveric) struck his interest, and although he had had no experience in osteology before he arrived in Cape Town, he published ten data-based publications on the topic by the time he resigned to take up the chair of anatomy in Durban in 1952.

The addition of Keen to the Cape Town department in the war years triggered a new interest in training research students. Writing in his annual research report for 1945, Drennan noted that the department was 'augmenting its collection of European, Coloured and Bantu skeletons from the material available each year in the dissecting rooms. The main object of this research is to make a statistical survey of the composition of and the racial characters of the coloured population.'[36] Until this point Drennan had not shown any interest in making a collection of cadaver skeletons and he had not done any detailed analysis of large samples of cadaver skeletons, so it is obvious that this was Keen's influence.[37] Drennan did publish one paper on the dentition of the cadaver sample (Drennan 1949), but the real work was Keen's (1950, 1951). Although Keen's studies were still intensively typological and relied on predetermined racial assessments, they are more systematic than the work of Drennan, even though the latter had over twenty years of anthropological publishing behind him. Singer remembers Keen as being much more helpful in learning research methods than Drennan.[38]

With Keen in the senior lectureship, Drennan was able to cope with the huge increase in medical anatomy students in the post-war years. Class sizes were now up to 300 students, and anatomy dissection was taught in two shifts: morning and afternoon.[39] Jack Keen's son, Ted, joined the department in 1947 as a junior lecturer just as Singer was completing his medical degree. Ted Keen had graduated in the medical class of 1943, but although Singer had shone in anatomy and this was directing his career choice, the younger Keen had returned to London after the war to train in surgery, but had become disillusioned, returning to Cape Town to work in anatomy as a second choice.[40]

Ted Keen was interested in clinical anatomy, but among his first publications was one on techniques of measuring cranial biometrics (Keen 1949) – something that his father had already developed some expertise in. This was followed with work on the estimation of stature from long bone lengths (Keen 1953), but purely anthropological papers were clearly not his primary focus. Singer, on the other hand, showed an interest in prehistory and anthropology right from the start of his career. Although not for degree purposes, Singer wrote an essay on the discoveries of fossil human remains in South Africa (Singer 1950). It was not original research, but a summary and interpretation of all the research on the topic in South Africa up to that point. The work won the annual Cornwall and York Prize for 1950. The prize had been presented by the University of the Cape of Good Hope from 1903 as a commendation for a scientific thesis or essay in order to stimulate original work among students of natural and physical sciences.[41] It is interesting to note that just six years after Singer's success in winning the prize, Sydney Brenner would be awarded the same prize for his thesis and go on to win the Nobel Prize in the future. In parallel with his nomination for the Cornwall and York Prize, Singer applied for a Rotary Foundation Fellowship and in this he was also successful. His fellowship would run for thirteen months in 1951 and 1952, during which time he would join the departments of embryology and biology at Johns Hopkins University in Baltimore to train in the methods of human and experimental embryology. This was a new research direction for Singer, but he continued his interest in anthropology by spending most of his weekends in Washington DC, where he gathered data on the skeletal variation at the Smithsonian Institution.

Singer returned to Cape Town in 1952 and proposed to Drennan that he refocus on embryology, but Drennan's response was not positive:

> I went to see Drennan, and I said, 'Look, I am not an experimental embryologist, but I am prepared to set up a lab and develop it here and to work hard and to train students.' 'Ag, Singer,' he said, 'that is absolutely impossible. We don't have the finances. In any event Singer you've got these wonderful studies going in anthropology and I think that is what you should be doing.'[42]

Singer was exceptionally disappointed and even though he tried to get senior university management to put pressure on Drennan to agree, Drennan would not relent and that was the end of his career as an experimental embryologist. Drennan's treatment of Ted Keen was not much better. Keen had put his foetal heart data together, which he had gathered during his forensic work, and submitted it as a doctoral thesis in 1952. Drennan refused to pass it on examination and an external examiner needed to be approached to evaluate the project. The external examiner was quite happy with the work and the dean passed the thesis.[43] Despite these difficulties, when Jack Keen left to take up the professorship in Natal in that same year, Ted Keen and Singer both applied for his senior lectureship position and Drennan had to decide which one to promote. Singer was the stronger candidate, but Keen was the son-in-law of Mrs Aimée Jolly, who sat on the university council and was a very good friend of Drennan's. Mrs Jolly was the widow of William Jolly, Drennan's late colleague in the medical faculty and former head of the Department of Physiology. In the end he solved the problem by getting the university to expand the post into two positions and Keen and Singer were both promoted.[44]

THE SALDANHA SKULL

Early in 1951, a fossil site on the farm Elandsfontein, near Hopefield, west of Saldanha Bay in the Western Cape, was brought to Drennan's attention.[45] Access to the site was difficult because it was actually a series of wind exposures in a sand dune field, but in May 1951 Singer

Figure 3.3: Donkey cart access to the Elandsfontein fossil site near Hopefield in the Saldanha Bay region of the Western Cape, *circa* 1952. There was no road access, even for four-wheel drive vehicles, at this time. (Source: Department of Human Biology, University of Cape Town)

Figure 3.4: Tea break during excavations at Elandsfontein in the early 1950s. From left to right: Keith Jolly (archaeology), Ronald Singer (anatomy) and Jack Mabbutt (geography). (Source: Department of Human Biology, University of Cape Town)

was able to get there via donkey cart and what he brought back to the department was stunning.

The fossils from Elandsfontein were of extinct forms of various mammals, but there were also Early and Middle Stone Age tools, providing confirmation of a human presence in these ancient times. John Goodwin, the UCT archaeologist, subsequently visited the site and not only confirmed Singer's observations, but also argued that the site was so valuable that there must be a formal archaeological investigation. At Goodwin's instigation, the university principal created a multidisciplinary departmental committee, consisting of anatomy, archaeology, geography, geology and zoology (Drennan 1953a, 480).

Among its early tasks was to hire Keith Jolly as university field officer to co-ordinate excavation efforts. Jolly, Drennan, Ted Keen and Jack Mabbutt of the geography department were regular visitors to the site in 1951 and 1952, and Singer joined them in fieldwork again on his return to South Africa.

On 9 January 1953, Ted Keen drove Singer and Jolly up to the site and left them there to begin that year's fieldwork, but because of a family obligation he did not stay.[46] Almost immediately, Jolly discovered fragments of a hominin in the deflation exposure he was examining, and shortly after Singer found more parts of the same skull nearby. This was the discovery of the famous Saldanha skull, but it also marked the beginning of a series of personality conflicts that would plague future research and publication.

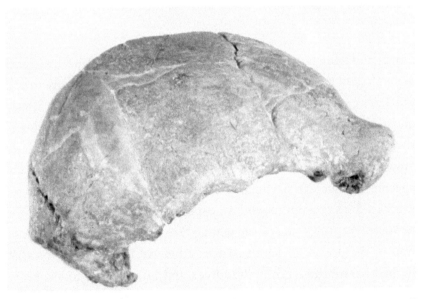

Figure 3.5: The Saldanha skull after reconstruction. The fossil site itself is known by the name of the farm, Elandsfontein, but the official designation of the fossil specimen is Saldanha 1. (Source: Iziko Museum)

Figure 3.6: E.N. (Ted) Keen holding stone and bone artefacts at the Elandsfontein fossil site, *circa* 1953. (Source: Department of Human Biology, University of Cape Town)

Singer and Keen brought the skull back to Drennan in Cape Town and were already making plans for the first publication on the discovery, but Drennan had other thoughts. In Ted Keen's words: 'We, Singer and I, decided we were going to study this skull but he [Drennan] then made it very clear that was not on his programme and *he* [Keen's emphasis] was going to study the skull, not us. He made it absolutely specific.'[47] Singer, in particular, was furious. Not only had Drennan sabotaged his attempt to work in experimental embryology, but now he was preventing Singer from getting the glory on what was the fossil find of the decade in South Africa. What bothered Singer was that he had been bringing him 'hundreds and hundreds of fossils every weekend', but when the skull was discovered, Drennan said, 'Singer, this is the answer to my dreams. ... I'd always been hoping to find Stellenbosch Man, and I think we have it now. I've always believed in it, and therefore I must describe the skull.'[48] Singer was especially bitter because he 'felt that here was an example of an old man trying to steal a bit of glory where he doesn't need it. He has enough. He is head of the department. He is recognised in Cape Town as being *the* [Singer's emphasis] anthropologist. In the newspapers he gets quoted.'

Indeed, Drennan did publish the first description of the cranium (Drennan 1953b). Drennan relented after this publication and gave Singer permission to publish his own paper (Singer 1954) and to join him in describing the mandible (Drennan and Singer 1955). Now Ted Keen was the aggrieved party, as Singer and he had agreed to acquiesce to Drennan's demand to describe the skull and leave the faunal material to the two of them. Keen published some of the faunal material (Singer and Keen 1956a, 1956b), but never forgave Singer for publishing on the human material and their personal relations soured. At this point Jolly began to express complaints that Singer was trying to steal the glory of the discovery from him. In fact, Drennan had been very careful to ensure that Jolly was noted as the discoverer in each of his papers, but Singer's 1954 paper had the two of them sharing equal glory. According to Singer, 'I have always said that Jolly found the first pieces, but I have also said that I also found some pieces and that they were all part of the critical picture, and therefore if we are going to talk about the discovery, I have to be a co-discoverer of some sort.'[49] The acrimony was loud enough that Goodwin asked Drennan for a formal clarification of the sequence of discovery. Drennan was concerned that the fragments were simply picked up from the ground and neither Singer nor Jolly had actually made any special search after the day of discovery, 'although they stayed on two more days at the site'.[50] They brought the fragments to Drennan on their return and presented them 'as some tit-bits they had found which Dr Keen had been non-committal about'. A week later Jolly, Singer, Mabbutt and Drennan went back to the site, but could find nothing more at the original site of discovery. That evening Singer produced a small fragment of bone

Figure 3.7: Matthew Drennan at the Elandsfontein fossil site, *circa* 1953. Although Keith Jolly was in charge of excavation at the site, Drennan, then in his late sixties, was a frequent visitor. He is holding a fossilised animal bone. (Source: Department of Human Biology, University of Cape Town)

that fitted perfectly with the pieces collected by Jolly. When Drennan asked, 'Where the h did you pick it up?', Singer said he had found it on his way home from the tent. The fact that this piece had come from some distance away from the original discovery presented a conundrum that bothered Drennan, as there was no easy explanation for how it could have been transported from one site to the other without human agency. Drennan told Goodwin that he 'would welcome any suggestion for a suitable explanation'. None was forthcoming.

According to his brother-in-law, Ted Keen, Jolly was 'a very difficult character' who never had an academic job despite being very clever.[51] As chair of the multi-department committee, Drennan had hired Jolly to run the fieldwork at Elandsfontein over the objections of Goodwin in the Department of Archaeology. The vice chancellor, T.B. Davie, had insisted that Jolly be part of archaeology, as placing him in anatomy would have prevented the university from finding funding for the archaeological fieldwork. Goodwin was extremely concerned that the quality of fieldwork at the site was not good enough, especially after the letter from Drennan describing the context of the discovery of the hominin skull fragments, but Goodwin had no effective control over Jolly since he was managed and paid by Drennan.[52]

Jolly had a degree in archaeology from Cambridge and had worked on the Peers Cave site on the Cape Peninsula in the 1940s, but his real claim to employment at the Elandsfontein site was his mother's presence on the UCT council. Relations with Goodwin were particularly frosty because Goodwin felt that Jolly's methods were not systematic enough and although Goodwin was nominally Jolly's supervisor, months would go by with Jolly neither submitting reports verbally nor in written form. Goodwin noted that Jolly rarely, if ever, recorded localities for the discovery of specific bones.[53] Singer goes so far as to accuse Jolly of salting the site with stone tools to give a better association between fauna and human activity.[54] Things were even more awkward for Goodwin because he was officially Jolly's Cambridge doctoral supervisor until Cambridge finally withdrew Jolly's registration in December 1952.[55] Eventually, Jolly's lack of professionalism and failure to advance the excavation convinced Drennan to fire him, but he was blocked by vice chancellor Davie.[56] Jolly became more and more difficult to work with and the complaints against him from Singer and others continued until he was finally released from the site in 1954.

Goodwin, who now had sole control of archaeological matters at the site, was becoming disenchanted with Drennan's management. Writing to Kenneth Oakley at the British Museum, Goodwin commented: 'I fear I distrust the medical approach, diagnosis from symptoms, with a vast proportion of intuition.'[57] Goodwin was also concerned because Drennan would not support detailed dating attempts, as 'they would merely muddle our deductions regarding evolutionary studies.'[58]

With Goodwin trusting neither Drennan nor Jolly, Singer and Keen unhappy about Drennan, Keen wary of Singer, and Jolly complaining about being overlooked and then fired, the relationships on the site were toxic and almost no solid archaeology with good contexts for the discoveries was being done. Things finally changed after the death of Goodwin and the hiring of Professor R.R. (Ray) Inskeep as head of archaeology at UCT in 1959. Ted Keen left to go to Natal, Drennan had retired, and Jolly had been removed. More importantly, Singer and Inskeep worked well together and the first real systematic and properly recorded excavations were conducted on the site (Singer and Wymer 1968).

So, what was Drennan's legacy and lasting influence at UCT? Perhaps Singer is not the best person to provide an opinion on this, but his views are telling. In an interview, Singer made it clear that he felt that Drennan was first and foremost a medical anatomist and it would be wrong to call him a biologist or physical anthropologist. According to Singer, Drennan's anthropological knowledge was elementary and his understanding of evolution was poor, but there were few in Cape Town with enough knowledge to refute him. Drennan 'was the authority on medical jurisprudence. He used to be called in on court cases, to identify skulls and stuff like that. Sure, he probably could identify some of them, but who knows the number of mistakes he made?'[59] Singer's bitterness is perhaps well deserved since there is no question that Drennan's primacy in the publication of the Elandsfontein discovery led to honours for him, which may not have happened if Drennan had ended his career in 1950, rather than 1956.

In the year of his retirement Drennan was elected a life member of the Anatomical Society of Great Britain and Ireland. The following year he was president of the South African Archaeological Society and received the South African Association for the Advancement of Science (S2A3) medal in 1957 and honorary life membership of S2A3 the following year. The University of the Witwatersrand (Wits) granted him an honorary Doctor of Science degree in 1958.

Whatever Singer's opinion, Drennan's long tenure in Cape Town, from 1913 until he finally retired, at the age of 71, in 1956, left an academic lifetime of collected materials in the department – from embryos and anatomical oddities to comparative anatomy specimens, human skeletons and fossils. Drennan's museum had no theme and was in reality a storehouse for interesting anatomical specimens, much like that of his alma mater in Edinburgh. For better or for worse, Drennan's collections have provided a wealth of information for future researchers. He did not arrive from Edinburgh with a pre-existing research agenda like his short-termed predecessor Thomson, but he did build a local South African focus as his interest in anthropology developed. He set a precedent for science research, at least in the 1920s and 1930s, and even though few of his science students continued in the field, the Department of Anatomy retained its postgraduate programme in anatomy long after Drennan's retirement. Drennan's research (even if it pre-empted work by Singer and Keen) put UCT on the map of fossil discoveries in South Africa. It could never be in the running against Dart at Wits, but it did mark UCT as a centre of anthropological research known around the world.

NOTES

1   Information on the Henderson Trust recorded in the anthropology catalogue in the Anatomy Museum at the University of Edinburgh.

2   There are no surviving copies of Vincent Vermooten's Master's thesis, but he did draft a paper from it, which he sent to Drennan. The draft is titled 'The

Long Bones of the South African Bushmen' and it has been stored in the Department of Human Biology at the University of Cape Town.

3   Iziko Museum, Cape Town, letter from Louis Péringuey to Frank Shrubsall, 3 January 1921.

4   Iziko Museum, Cape Town, letter from Péringuey to Shrubsall, 29 April 1921.

5   Vermooten describes the vertebrae from four executed criminals whose bodies were sent to the University of Cape Town in 1919 for dissection. Charles Slater (1999, 258) has described the cadaver of an executed criminal from the first dissection at the Medical School in 1911, but he believes that this was the only case where such bodies were used. Vermooten's paper indicates that this happened more often, but there are no other records of executed criminals being used for medical school dissection in South Africa.

6   Department of Human Biology, University of Cape Town, letter from Vincent Vermooten to Matthew Drennan, 15 October 1921.

7   Department of Human Biology, University of Cape Town, letter from Vermooten to Drennan, 15 October 1921.

8   Department of Human Biology, University of Cape Town, letter from Matthew Drennan to Eugène Pittard, 1 November 1927.

9   Cadaver number 285 is listed as a woman who died at Somerset Hospital on 29 June 1919. The cadaver records list her race as British.

10   Department of Human Biology, University of Cape Town, Vincent Vermooten, 'A Note of a Dissection of a Female Hottentot', 1919.

11   The first catalogue of human skeletons in the department was drawn up by Drennan on 13 August 1925. The specimens had been unnumbered prior to that.

12   UCT 80 in the skeleton collection is this specimen. He remains anonymous because no cadaver number was kept that would enable us to link him to the records of the cadaver intake of 1925.

13   Department of Human Biology, University of Cape Town, letter from Matthew Drennan to J.C.B. Grant, 15 February 1927. The Grant collection skeletons moved from Winnipeg to Toronto with Professor Grant and the presence of the original 'Bushman' skeletons was noted there by Professor Susan Pfeiffer. Pfeiffer and I, in Cape Town, exchanged the specimens in late 2012 and early 2013. The South African specimens were re-accessioned at the University of Cape Town with the other archaeological individuals, and the Canadian specimen was forwarded to Professor Rob Hoffa at the Department of Anthropology at the University of Manitoba. Since the Canadian specimen was of cadaver origin, Hoffa proposed that an attempt should be made to identify the ethnicity of the individual and his remains possibly returned to the descendant community.

14   Department of Human Biology, University of Cape Town, Drennan correspondence about embryo donations.

15   Department of Human Biology, University of Cape Town, Sam Scher, 'Description of an Early Human Embryo-Ovum'.

16   Although these donations were initiated by the clinicians, rather than by Drennan himself, a few of the donors were well known to him. Dr G.B.

Wilkinson, a doctor in Sea Point, actually added a bit of local medical gossip to the back of his letter of donation. The scrawled note asks Drennan if he had heard 'that our hairy and argumentative friend from Potchestroom – Kramer – has left his wife and she is suing him for restitution. I'd pay him to keep away rather' (Department of Human Biology, University of Cape Town, Drennan correspondence about embryo donations).

17   Department of Human Biology, University of Cape Town, letter from Arthur Robinson to Matthew Drennan, 1 August 1918.

18   Department of Human Biology, University of Cape Town, letter from Matthew Drennan to Arthur Keith, 23 May 1929.

19   Department of Human Biology, University of Cape Town, letter from Dorothea Bleek to Matthew Drennan, 24 April 1924.

20   Solly Zuckerman was disparaging about his experience in the department and was determined to leave South Africa as soon as possible. He was underwhelmed by Drennan and complained about his focus on rote learning of anatomical detail and lack of research supervision (Zuckerman 1970, 74; 1978, 12). It is unclear if Zuckerman's problem was with this education, or the fact that he felt that South Africa was too parochial and no place for a man of his talents (Morris 2009b, 238).

21   Phillip Tobias (1987, 32) has argued that Dart's collection of cadaver skeletons can be linked back to Sir William Turner in Edinburgh, through Robert J. Terry at Washington University in St. Louis in the United States. According to Tobias, Terry was inspired by Turner's collection in the 1890s and began his own collection of cadaver skeletons when he returned home. In turn, Dart was inspired by Terry – hence the genesis of the Dart collection of human skeletons in Johannesburg. My opinion differs, as it is Drennan's collection in Cape Town that is most similar to Turner's in Edinburgh and he did not begin to systematically keep cadaver skeletons until the 1940s. Terry's idea of a records-based cadaver collection was his own.

22   Iziko Museum, Cape Town, letter from L. Kemper enclosed with note from Colesberg Magistrate to Gill at South African Museum, 7 December 1927.

23   Department of Human Biology, University of Cape Town, note from James Drury to David Slome, undated.

24   More recent visions of the work (Peckmann 2002) mark the cemetery as that of a typical poor community on the margin of the country town, showing the dynamics of ethnic transformation in nineteenth-century rural South Africa, which includes a wide range of biological admixture.

25   Note in University of Cape Town Department of Anatomy file, dated 29 July 1954, signed by Ronald Singer, acknowledging borrowing the files on Cape Flats, Kalk Bay, Fish Hoek and Plettenberg Bay (Pedomorphism). Note added on 1 September 1955 to include the boxes: Geological Survey Southern Rhodesia, Likasi Skeleton and Oakhurst.

26   Singer, with his access to Drennan's original notes and correspondence, published an extremely critical note on the Cape Flats discovery in 1993. He quite correctly noted that all of the discoveries were mixed and clearly conflated in windblown sand deposits. Very obvious Dutch colonial clay

bricks were mixed with Later Stone Age tools and fragments of one of the associated skeletons produced a modern radiocarbon date. The Cape Flats skull itself, although large and robust, fits within the range of variation of modern Khoesan populations.

27 Department of Human Biology, University of Cape Town, letter from Keith to Drennan, 6 October 1926.

28 Department of Human Biology, University of Cape Town, letter from Keith to Drennan, 18 March 1929.

29 Department of Human Biology, University of Cape Town, letter from Drennan to Keith, 3 May 1929.

30 Isidore Zieve graduated with his Master's in anthropology in 1929. There is no record of his project, but Drennan does mention in his publication (Drennan 1929a) that Zieve was responsible for measuring the Cape Flats skull, so we can assume that this work was part of his thesis.

31 Royal College of Surgeons of England Archives, London, letter from Matthew Drennan to Arthur Keith, 19 December 1929.

32 Department of Human Biology, University of Cape Town, letter from Matthew Drennan to Jack Heselson, 20 August 1937.

33 Interview with Ted Keen, 6 December 1994, Department of Anatomy, University of Cape Town.

34 Interview with Ronald Singer, 20 November 1991, University of Chicago. Keen's publication record shows that he published four original papers in Portuguese between 1945 and 1948.

35 Interview, Keen.

36 Department of Human Biology, University of Cape Town, 'Department of Anatomy Research Report for 1945'.

37 There is also a possible connection to Coert Smit Grobbelaar from the University of Stellenbosch who studied anthropology with Drennan in the years 1939–1940. The Department of Zoology at the University of Stellenbosch had a particular interest in the biology of coloured people. See Chapter 4 in this volume.

38 Interview, Singer.

39 Interview with Hertha de Villiers, 7 February 1990, Parkhurst, Johannesburg.

40 Interview, Keen. Ted Keen had started medicine at Charing Cross in 1937, but decided to join his father in Cape Town when the class was evacuated to Glasgow because of the bombings in 1940. He was given credit for two years of his United Kingdom medical degree and completed the course at the University of Cape Town.

41 University of South Africa (Unisa) Institutional Repository (online at http://uir.unisa.ac.za), Cape of Good Hope, Colonial Secretary's Ministerial Division, 'Report of the Council of the University of the Cape of Good Hope for the Year 1901'. http://uir.unisa.ac.za/bitstream/handle/10500/6332/UCGH_R1901.pdf? sequence=1&isAllowed=y, accessed 12 July 2021.

42 Interview, Singer.

43 Interview, Keen.

44 Interview, Singer.

45 The fossil hominin cranium that was recovered from the site is correctly labelled Saldanha 1, but the names 'Elandsfontein' and 'Hopefield' are often used as synonyms.

46 Interview, Keen.

47 Interview, Keen.

48 Interview, Singer.

49 Interview, Singer.

50 University of Cape Town (UCT) Special Collections, Cape Town, letter from Matthew Drennan to John Goodwin, 30 March 1953.

51 Interview, Keen.

52 UCT Special Collections, Cape Town, letter from John Goodwin to Clarence van Riet Lowe, 19 August 1952.

53 UCT Special Collections, Cape Town, letter from John Goodwin to Kenneth Oakley, 30 August 1952.

54 Interview, Singer.

55 UCT Special Collections, Cape Town, letter from Cambridge Board of Research Studies to John Goodwin, 4 December 1953.

56 UCT Special Collections, Cape Town, letter from Goodwin to van Riet Lowe, 19 August 1952.

57 UCT Special Collections, Cape Town, letter from Goodwin to Oakley, 30 August 1952.

58 UCT Special Collections, Cape Town, letter from John Goodwin to Berry Malan, 24 April 1957.

59 Interview, Singer.

4

The Age of Racial Typology in South Africa

The South African scientists studying the bodies and bones of living and extinct people in the first half of the twentieth century were working with a scientific philosophy firmly rooted in the work of Carl Linnaeus done two centuries earlier. Linnaeus, the father of systematic biology, spent his entire scientific career attempting to name and describe all of God's creation. He felt that his labours would reveal the Creator's plan in the Great Chain of Being. Each link of the chain was represented by a separate species, and the arrangement of the species on the chain, from simple to complex, represented God's plan as revealed by anatomy (Lindroth 1983, 24).

Linnaean classification required a precise and fixed definition of each species so that this order could be uncovered. Each species had to be invariable and unchanging for Linnaeus's system to make sense (Lindroth 1983, 24). Linnaeus began in his later years to accept that new varieties could and did appear, but he downplayed their importance. Linnaeus felt these deviations were at odds with nature, and in particular with the nature of the species (Eriksson 1983).

The Linnaean vision of fixed species was an outgrowth of the philosophy of Plato. The Platonic view considered that each species had an eternal ideal or essence. This was the archetype, or original form of the species. Linnaeus adapted this to his concept of a species in which he 'considered that the archetype was the most important measure of a species' (Magner 1979, 353). To describe a species was to describe its archetype and variations were only shadows that departed from the original pure ideal.

In his first edition of *Systema Naturae*, Linnaeus catalogued human races as full species in their own right, but by the tenth edition he had transformed the races of humanity into varieties. But these races did not fit his standard approach to varieties. Each was provided with a brief description of its essence and its own name in Latin as if it were still regarded as a separate species. Linnaeus clearly intended these human variants to be considered fundamental units in the order of nature.

The Linnaean concepts of classification and archetype found fertile ground in the natural sciences and, particularly, they became firmly rooted in the young field of anthropology. The technique of racial description that developed in this era became known as *typology*. It depended on a definition of the type in which all the important and fundamental characters of the race were seen. Hence the focus of racial classification was only on those features that could differentiate between races. A pure race would be one that had not intermixed with other races and whose component individuals were still exactly like the original type, the archetype. Since the type was an ideal standard, any individual could be compared to the type and his or her purity assessed. Variations were seen as impurities, but the characteristics of the type could be dissected out through careful observation. The task of the typologist was to explore the stable essence in each individual, however much these features were disguised by variation (Stepan 1982, 138).

From its inception, racial typology in anthropology crossed the bounds of dispassionate science to become a technique of racism. Throughout the nineteenth century, craniology laboured to identify the controlling factor or essence that marked the Europeans as superior to all other human varieties. Since the seat of intelligence was the brain, the size of the brain (as reflected by cranial capacity) was taken to signify intelligence. Careful selection of type specimens enabled anthropologists of European origin to prove that Europeans had the largest brains and were therefore the most intelligent of all human races (Gould 1978, 508). This was why the exceptionally large cranial capacity of the Boskop skull was of such importance and why researchers such as William Pycraft (1925, 191) argued that even if Boskop was exceptional, it was still inferior to the Cro-Magnons of the European Upper Palaeolithic. Craniological studies in Europe took on political overtones and focused on the issue of national types. Individuals were sorted according to their morphology into distinct types so that each European nation could discuss the racial background of its citizens. Courses in physical anthropology taught to medical students such as Matthew Drennan, Raymond Dart and Frederick FitzSimons used craniology to reconstruct the various pure types that had fused to create the modern racial complexities. Although the technique was based on physical measurement, it extended into the belief that mental abilities and national character were determined by race. This was the message underlying much of the anthropology presented by Alfred Cort Haddon and Felix von Luschan in their papers to the South African Association for the Advancement of Science (S2A3) meeting in South Africa in 1905.

Both Drennan and Dart arrived in South Africa long after that S2A3 meeting, but there were others present at the meeting who would also influence the field of physical anthropology in future years. Robert

Broom had arrived in South Africa in 1897 and had spent most of the succeeding years trying to set up a medical practice in order to earn a living. In 1903 he obtained the post of professor of zoology and geology at Victoria College in Stellenbosch and he remained in that position until 1910. Thomas Dreyer and Coert Smit Grobbelaar were both students at the time of the S2A3 meeting in 1905. Both were top scholars, Dreyer graduating in 1905 with a first-class pass in zoology under Professor Arthur Dendy at the South African College in Cape Town, while Grobbelaar was a first-year student under Broom at Victoria College in 1905. It is almost inconceivable that these two excellent students were unaware of the scientific meeting happening in Cape Town. Sadly, although it is possible to identify delegates who spoke in the scientific sessions, there are no published records of attendees. Both of their professors were on the local organising committee for the conference and Broom almost certainly must have talked about the conference to his students. George Findlay (1972, 25) described Broom's lectures while he was at Victoria College as being especially lively when he strayed from the subject under discussion to express his views on topics of wider concern. The two students joined the S2A3 once they had their professional degrees: Dreyer in 1915 on his return from doctoral studies in Germany, and Grobbelaar in 1917 after he was hired as a lecturer in zoology at Victoria College.

Saul Dubow (1995, 35) has emphasised that physical anthropology took the lead in the study of South African prehistory after the conference in 1905. Much of this was the direct result of the influence of Haddon and Von Luschan in their recommendations to study the dying races and especially in relation to the study of the prehistory of these people before the miscegenation of modern times. Broom, Dreyer and Grobbelaar would all become major actors in this investigation. Typology was at the heart of this kind of physical anthropological study, and it would result in a new and more directed phase of research, with a focus on racial studies in the 1920s and 1930s.

Elsewhere, I have bounded this period by the discovery of Boskop in 1913 and its debunking in 1958 and referred to it as the 'Age of Typology' (Morris 2005b, 129; 2012, S157). Although I use Boskop to define the age, Boskop was just one of a series of types that were constructed to give shape to the origins of the living peoples of South Africa.

THOMAS FREDERICK DREYER (1885–1954)

Dreyer, like Louis Péringuey, was an entomologist who moved into anthropology. One of six boys, Dreyer was born on a farm at Koeberg in the Western Cape. When his father died young, his mother moved the family to her sister's place in Stellenbosch, where he grew up. Marie Barry, Dreyer's daughter, was told by her father that he had wanted to join the Boer forces in the Free State Republic when the South African War broke out in 1899, but he was too young.[1] Dreyer attended the South African College (University of Cape Town [UCT] after 1918) and was the first student to graduate under the first chair of zoology, Arthur Dendy, in 1905 with a first-class Bachelor of Arts degree (Brown 2003, 89). Upon graduation, Dreyer took a job with the Department of Agriculture to be trained in entomology for plant inspection. In 1907 the department arranged a Queen Victoria scholarship for him to pursue advanced zoological studies in Germany (Lounsbury 1940, 20).

Dreyer's choice of graduate school, the University of Halle-Wittenberg (known as Martin Luther University after 1933) was a logical choice not only because it specialised in agricultural topics, but also because several other South Africans were already studying there. Germany was a popular choice for Dutch-speaking South Africans and Dreyer told his grandson that his years in Halle were the best times of his life.[2] Dreyer kept a book of photographic remembrances signed 'Thos.F.Dreyer, Halle 1908', which was passed down to his daughter. Dreyer was active in student life and took up fencing while in Halle,

Figure 4.1: Thomas Frederick Dreyer. This photograph was taken as a publicity shot for a public lecture at the National Museum in 1937. Dreyer was in his mid-fifties and at the height of his academic career. (Source: National Museum, Bloemfontein)

sporting a faint duelling scar below his moustache from a misdirected point. Although Afrikaans was his mother tongue and English was the language he preferred to publish in, Dreyer told his grandson that German was his favourite language. Dreyer graduated with a doctorate from Halle in 1910 and returned to South Africa via the United States (where he attended courses at Cornell University in New York) and Australia, where he visited several agricultural institutes (Van Pletzen 1987). Upon his return, he was appointed assistant entomologist with the Department of Agriculture and based in Oudtshoorn. He resigned his post in 1912 to take up a lectureship at Grey University College in Bloemfontein and was promoted to professor at the beginning of 1917 (Plug 2020a).

Dreyer's publication record at this time was quite respectable. Between 1907 and 1919 he published twelve papers on a mix of entomological and zoological topics. One of his papers was in German, but all of the rest, except for a popular Afrikaans pamphlet on the animal kingdom, were in English. Then his publication record stops for twelve years, only to reappear in a different guise after 1931. From that date Dreyer published another twenty papers, of which only one was on a zoological topic. All of his other post-1931 publications were on anthropology or archaeology. What happened between 1919 and 1931 to stop Dreyer from publishing and how did his complete shift from biology to anthropology and archaeology come about?

There are a few clues. We know that Dreyer was an active teacher of undergraduates and that he was very engaged with the college senate after his promotion in 1917. Since Grey University College fell under the administration of the University of South Africa at the time, Dreyer's senate activities extended to the chairmanship of the senate of the parent body as well (Hoffman 1955b, 186). Dreyer was a member of the S2A3 and was on the local organising committee for the 1923 congress in Bloemfontein, but he does not appear to have taken any role on the S2A3 council as a member for the Orange Free State. Perhaps this was a result of his more intense engagement with Grey University College at that time. Dreyer's name was proposed for rector in 1927, but the committee did not offer him the post. One of the reasons was that he was too English-speaking and gave his lectures in English (Van der Bank 1995, 345). Dreyer must have been disappointed, but what was worse was that he was also having personal problems around this time. Dreyer and his wife were struggling with their marriage and she had left Bloemfontein to stay in Oudtshoorn, where she was originally from.[3] Dreyer was now maintaining homes in both locations and he would spend long periods in Oudtshoorn trying to woo her back. Ultimately, he was successful, and she returned to their Bloemfontein home, but their daughter Marie, who would have been a teenager at the time, had fond memories of living in Oudtshoorn between 1926 and 1929.[4] It is exactly during these years that Dreyer began to excavate the nearby archaeological site at Matjes River.

Dreyer's shift to archaeology was absolute. Albert (Bertie) John Dirk Meiring was Dreyer's student in the mid-1920s and returned to the Department of Zoology at Grey University College as a lecturer in 1934. Dreyer was working on migratory red locusts when Meiring left Grey University College in 1925, but when he returned in 1934 all of Dreyer's research was in archaeology or anthropology.[5] There are two hints that Dreyer's interests were changing earlier than this. Dreyer corresponded with Péringuey at the South African Museum in 1915 about a heavily mineralised human mandible that had been discovered near Harrismith in the Orange Free State. Dreyer had prepared some notes for publication on this discovery and had sent both the specimen and his observations to Péringuey for comment and guidance. Dreyer suggested that the find was of great antiquity and could be compared to what was then the new discovery at Boskop, but Péringuey was gently dismissive, suggesting that its preservation was similar to many other examples of 'Bushmen' held in the collection in Cape Town. Péringuey suggested that Dreyer should go to the South African Museum and study the comparative material that was available.[6] Dreyer did not take Péringuey up on his offer, but although his interest in these archaeological human remains may have been dampened, it was not extinguished. In 1917 he walked from Bloemfontein to the spa at Florisbad and was intrigued by the fossilised bones that the owner of the spa showed him. This was Dreyer's first contact with the site that would make him world famous in the 1930s (Hoffman 1955a, 166). His hiking partner on the 1917 trip was D.F. Malherbe, the first chair of Afrikaans at Grey University

College and the person who would vote against him in his application to become rector of the College because he was too English.[7]

But the real transition from zoology to archaeology happened when Dreyer discovered the archaeological deposits and their human inhumations at Matjes River. Louw (1960, 15) tells us: 'It was coincidence which led Dreyer to the discovery of the cave in 1928', but nothing of the context of discovery is mentioned in any of Dreyer's published papers. Dreyer did release information on the site in the form of an interview in a newspaper article in 1930.[8] The article tells us that some months before Dreyer had obtained from Professor W. von Bonde of Grey University College some skull fragments from a cave at the mouth of the Matjes River, Plettenberg Bay. It goes on to say that Dreyer spent several weeks excavating the site. We do know that Dreyer received funding in 1928 for his initial excavation from the Research Grant Board of South Africa, so if we extend the 'some months before to perhaps eighteen months, then this makes sense. What we also know is that Dreyer extended his initial excavations in 1929 and by 1930 he had not only accumulated a large number of archaeological artefacts, but also more than 30 human burials.

Dreyer had no formal training in archaeology. How he taught himself field archaeology is unknown, but much of the terminology he used came directly out of the European archaeological literature, suggesting that his knowledge came from reading the published sources. Added to that is his correspondence with other local researchers who were very much active, but for the most part also without training. Two of these were John Hewitt of the Albany Museum in what was then Grahamstown and Drennan in Cape Town. Hewitt was himself an amateur archaeologist, but he had worked closely with other amateurs (there were no professional archaeologists in South Africa before the appointment of John Goodwin to UCT in 1923) and had gained a great deal of experience in excavating caves in the Eastern

Cape (Binneman 1990, 14). It was Hewitt who had coined the term 'Wilton Culture' (Hewitt 1921, 460). As his discoveries at Matjes River accumulated, Dreyer wrote to Hewitt for help in identifying stone cultures from tools. Dreyer wrote to Hewitt while he was busy trying to write up his discoveries in September of 1930: 'I sent you a small box of stone implements from Matjes River Cave some days ago in the hope that you may be able to "place" them – which is more than I can do definitely.'[9] Further on in his letter he noted that it was not only the Matjes River implements that were puzzling him. He told Hewitt that he had three other collections – from Rooiwal on the edge of the Kalahari; from Douglas on the Orange River, in association with very ancient graves; and a newly discovered site from a high terrace on the Kamassie River near Oudtshoorn. The Oudtshoorn site had 'cores (bouchers) and flakes, scrapers, knives, etc.' and he told Hewitt that he was planning to go back to the site again soon.

Drennan was Dreyer's correspondent of choice as he began to work on the physical anthropology of skeletal discoveries. While working on the first publications in 1930, Dreyer wrote to Drennan requesting comparative anatomical details from Drennan's published works, but also his opinion on the works of others.[10] The two researchers kept up a correspondence between 1930 and 1938, and Drennan made a point of visiting Dreyer in 1931 to see the material excavated from Matjes River. Drennan complimented Dreyer's work, saying that he had produced 'an outstanding collection of skulls of the pre-Bushman type, containing material which was unique'.[11] Drennan's correspondence with Dreyer became more focused on fossil remains when Dreyer announced the discovery of his fossil human at Florisbad.[12] Dreyer had been recovering non-hominid fossils from the site for some time, but on 26 July 1932 the eye of the spring at Florisbad produced the calvarium and partial face to which Dreyer would give the name *Homo helmi* in honour of the donor who was helping to fund his excavations (Henderson 1992, 18; 1994, 28).

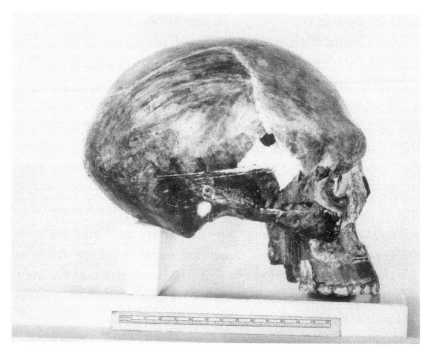

Figure 4.2: Thomas Dreyer's reconstruction of the Florisbad hominin. The skull was found in several pieces. Dreyer's reconstruction had several anatomical flaws, which made it appear more modern. The reconstruction was revised in the 1980s and the skull is now considered to represent a form more ancient than modern humans. (Source: Department of Human Biology, University of Cape Town)

The discovery of 'Florisbad Man' would bring Dreyer international notoriety and would also cement his reputation as an archaeologist. In fact, it is evident from his writings that his field techniques were limited and the recording of archaeological data and context poor. Although he went to some effort to describe the human remains recovered, subsequent researchers have struggled to confirm exact locations of his discoveries in the stratigraphy. Ericka L'Abbé, Marius Loots and Natalie Keough (2008, 61) unpacked the entire assembly of Matjes River skeletons from the National Museum collection (including all of Dreyer's excavations and the additional excavations done by Abraham

Hoffman and Meiring in the 1950s). They located 172 boxes stored from the site, of which only 76 (44 per cent) housed single individuals. All of the other boxes had comingling of skeletons – which may have been caused by the state in which they were excavated or resulted from poor curation afterwards. More importantly, 83 boxes (nearly half) had no archaeological provenance at all and most of the material had been sorted into boxes by skeletal element. Dreyer published no detailed description of the excavation site to reveal the exact location of burials and showed little interest in skeletons that he perceived to be 'Bushman' (that is to say, more modern).

Despite this lack of rigour in excavation technique and recording, Dreyer was becoming recognised in South Africa as an archaeologist of note. When the Inter-University Committee for African Studies was launched in 1933, one of its first acts was to create a subcommittee to examine the status of prehistoric studies (Goodwin 1935, 291). The members chosen were Mrs W. Hoernlé (University of the Witwatersrand, social anthropology), Dr J. Engelbrecht (Bantu languages, University of Pretoria) and Professor T.F. Dreyer (Grey University College), under the chair of John Goodwin (UCT). Dreyer was in esteemed company when Professor Clarence van Riet Lowe was co-opted onto the committee the following year. Goodwin and Van Riet Lowe were the only two professional archaeologists in the country at that time.

By 1934, Dreyer was confident enough to embark on a major project – a full monograph on the archaeology of South Africa. The following year he submitted a manuscript titled 'The Evolution of the Palaeolithic Culture in South Africa' to the University of South Africa for publication, but to his great displeasure the volume was not accepted after criticism from three reviewers. Dreyer was disheartened and filed the manuscript away until it was rediscovered in the papers of the National Museum in 1981. Meiring, Dreyer's student, then well into his own retirement, proposed that the manuscript should be published for its historical value and he wrote a short foreword for the proposed volume, but again nothing came of this and the manuscript still sits in the archives of the National Museum in Bloemfontein.[13] So, what did Dreyer say and why was the book rejected?

The manuscript is just more than 100 pages in length. The arch-aeological sites are all Dreyer's explorations, and the focus is solely on the stratigraphy and stone tool typology of the discoveries. Most important is that Dreyer expends significant effort in creating what is effectively his own stratigraphic terminology, comparing it to European standards and loosely linking it to the work done earlier by Goodwin and Van Riet Lowe. To Dreyer, archaeological strata were defined by the associated stone tools, which were linked to specific people as defined by ethnicity or race. Dreyer did not consider that different layers might reflect different activities of the same people. This is apparently the issue that upset the reviewers. One reviewer lamented that nearly the whole discussion was based on stratigraphic data alone, but Dreyer's conclusions 'nearly all are open to attack upon examination as being either incomplete or too particular for gener-alisation'. Another reviewer was even blunter: Dreyer's manuscript 'suggests very considerable enthusiasm, but rather hurried field-work neither of which are backed by an adequate knowledge of his subject'.[14] The net result of these criticisms is that Dreyer began to move away from field archaeology to focus more on the physical anthropology of his skeletons.

COLLECTING SKELETONS IN THE KAROO

An underlying theme in all of Dreyer's research at Matjes River was the idea that each consecutive level of stratigraphy, with its attendant stone technology, was linked to a specific human type. He, like most of the researchers at the time, saw South African prehistory as a series of migrations and replacements, of which the European colonisation was the most recent. Thus, there was a fascination with the description of prehistoric types and a concerted effort to describe each of these types in craniological detail. Much of this typological research concerned itself with description of Khoesan peoples who were believed to be on the edge of extinction. Key researchers in the early years were the Johannesburg scientists Dart, Lawrence Wells and Alexander Galloway, but also Drennan in Cape Town, Broom and of course Dreyer in Bloemfontein. All published descriptive papers on the craniology of 'Bushmen' and 'Hottentots', both living and archaeological. But this frenzy of typological assessment also contributed to what became a perfect storm of skeleton collecting in the vast inland area of the Karoo and the adjacent grasslands to the north and east. The collection of archaeological skeletons in South Africa's dry hinterland was an easy exercise for both amateur and professional scientists during the age of typology because the dry climate meant that early historic populations of pastoralists had sought out the river systems to water their flocks and bury their dead. Added to this was the cultural choice of using stone cairns to mark graves, so these early burials were highly visible on the landscape in riverine contexts.

The earliest collector of skeletons from this region was a young man barely out of his teens. Vernon Hofmeyr Brink travelled through the district of Griqualand West (south and west of Kimberley) in 1919 and 1920, stopping at local farms and excavating more than 50 graves, which the farmers identified as belonging to historic Griqua people, and 27 skeletons were collected (Brink 1923, 148–149; 1933, 15). Brink wanted to look at Griqua people because historical accounts identified them as a first-generation cross between Europeans and Khoekhoe – 'Hottentots' in the language of the day. Brink was born in Kimberley in 1898, the third child of Arend Brink, who was the senior diamond valuer for De Beers Company.[15] His father's wealth and social stature would have provided Brink with a privileged childhood and funded his education first at Victoria College/University of Stellenbosch, and then at Oxford. His mother, Maria Odendaal, came from an old Kimberley farming family and this would have given young Brink his connection to speak to the local white farmers about the presence of burials on their properties. Brink enrolled at Victoria College in 1916, studying English, Dutch, physics, chemistry and zoology. He graduated with a Bachelor of Science degree in 1918 and stayed on to complete a Master's the following year. Both degrees were passed *cum laude*.[16] When the influenza epidemic struck in his last year of study Brink volunteered to help his fellow students.[17] Brink gives us no clue as to how the skeletons he excavated were part of his degree, but it seems likely that they were the focus of his Master's investigations. Brink left South Africa in 1920 to study medicine at Oxford and arranged for his skeletons to follow him to England, so that he could

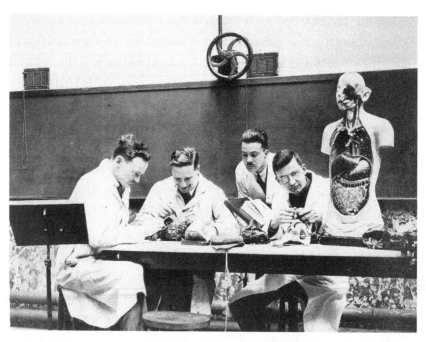

Figure 4.3: Vernon Brink (third from left) with other unnamed medical students studying anatomy from models at Oxford University, *circa* 1921. (Source: Dale Fischer, with thanks to David Morris)

continue to work on them (Brink 1933, i). His first paper was written as he began his clinical studies in London, but it appears he was essentially self-taught, with limited supervision. Writing to the professor of anatomy at Oxford, who had been storing the skeletons and providing workspace to study them, Brink notes: 'I fully realise the great incompleteness of the paper – it is rather an amateurish product of the phase before I had even commenced any study of anatomy'.[18] Brink continued to work on the data he had gathered from the bones, resulting in a doctorate from the University of Stellenbosch in 1930, but sadly the actual skeletons were no longer available to him as they had been loaned to Eugène Pittard in Geneva. He kept tabs on the specimens, with the intention of eventually continuing his study, but in the end never completed the description of the skeletons, and neither did he do any further excavations.[19]

At the same time that Brink was mulling over his data in London, another excavation on a much grander scale began outside the Karoo town of Colesberg. The South African Museum in Cape Town received news of a cemetery of 'Bushmen' who had died in the great smallpox epidemic of 1866, and the museum taxidermist James Drury was assigned their excavation as part of the museum's programme of accessioning San skeletons. Although Drury had no formal training as an archaeologist, he had already honed his excavation skills through earlier work at Coldstream Cave. Drury was meticulous as a fieldworker, but he could not analyse the skeletons as he had little actual knowledge of physical anthropology. Drury asked Drennan for help. Drennan insisted on separating out those skeletons with distinct Khoesan features and lumping them together as a 'Bushman Tribe' (Drennan 1929b, 62). This was consistent with Drennan's focus on pure races that he was beginning to articulate in Cape Town. Drennan held this position despite Drury's and David Slome's observation that there was a lot of variation both in grave style and morphology of the skeletons (Slome 1929, 59). Drury was very aware of multiple ethnicities in the cemetery, writing: 'My own personal opinion is that these 55 graves contained a mixed lot of natives some of which are good Bush types, and not unlike some of the Kalahari, also Strandloopers. I also think that there is a number of Bantus ... also Hottentot and possibly N°.29 is a bastard European.'[20]

ARCHAEOLOGY AND PHYSICAL ANTHROPOLOGY AT THE NATIONAL MUSEUM IN BLOEMFONTEIN

The skeletons excavated by Drury and Brink were accessioned to the South African Museum in Cape Town, but the two regional museums, the National Museum in Bloemfontein and the McGregor Museum in Kimberley, were also accumulating large numbers of human skeletons (Morris 1992a, 5–6). Dreyer's academic post was at Grey University College, but throughout his career he worked closely with the National Museum in Bloemfontein. Not only did the museum provide him with workspace and storage for his excavated specimens, but it also provided a venue where archaeology was a subject of study and discussion. The director of the National Museum for much of Dreyer's time was Egbert Cornelis Nicolaas van Hoepen (1884–1966). Van Hoepen had emigrated to South Africa as a child and at a young age volunteered for the Boer forces in the South African War. Captured by the British in 1900, he was deported to his home country of the Netherlands, where he later qualified as a mining engineer and then competed a doctorate in 1908. His thesis was of such a high standard that he 'qualified simultaneously as a geologist, mineralogist, and palaeontologist' (Badenhorst 1968, 700). He returned to South Africa as a palaeontologist, first taking employment at the Transvaal Museum in Pretoria and then being appointed as director of the National Museum in Bloemfontein in 1922.

The stone tool archaeology of the Free State became a special interest of Van Hoepen's and he published three papers on the topic between 1926 and 1928. His last paper (Van Hoepen 1928) is particularly important because he used it to launch a series of archaeological research papers at the museum under the title *Argeologiese Navorsing van die Nasionale Museum* (Archaeological Research of the National Museum). Although his training was in palaeontology, not archaeology, Van Hoepen declared his own expertise in the subject and after the death of Péringuey in 1924, he went so far as to demand that the South African Museums Association authorise the transfer of all archaeological collections in Cape Town to his curation in Bloemfontein (Mason 1989, 77). The association ignored his request, but that did not stop his continued interest in archaeology. Starting in 1932, Van Hoepen published a series of papers in the *Argeologiese Navorsing van die Nasionale Museum* on stone cultures along the Cape south coast and extended this into the Transvaal after Hoffman joined the museum staff in 1934.

Hoffman (1903–1969) was a Stellenbosch-trained zoologist who had not worked in archaeology before his arrival in Bloemfontein, but he rapidly developed an interest in the subject and began to excavate human skeletons on his own in the 1940s. In October 1934, Hoffman and Van Hoepen and three labourers travelled to Buispoort near Zeerust with the express purpose of investigating 'old mines, graves, beads, etc.'[21] Although only a small number of human bone fragments were recovered, Hoffman and Van Hoepen collected archaeological and ethnographic artefacts of the Bahurutsi people. The results of the expedition were published in the museum's *Navorsinge* (Van Hoepen and Hoffman 1935). Hoffman, with two workers from the museum, travelled to the town of Philippolis in September 1941 with the express purpose of excavating Griqua skeletons.[22] Like Brink before him, Hoffman was specifically interested in this group because of the question of the intermixture of races. The focus on specimens for display as racial types becomes obvious when we read Hoffman's words from the report: 'Thirty graves were excavated; 27 adults and three children. Only one or two skeletons are complete and suitable for mounting.' Hoffman made another expedition within a year to collect another set of type skeletons, this time 'Bantu'. The Hoffman family farm at Kareeboom in the Wolmaransstad district of the Transvaal presented a series of 'Bantu' graves that Hoffman's father wished to remove in order to cultivate the land in which they were located. In March 1942 Hoffman and two labourers from the farm began to excavate, but had to stop when 'complaints began to come in from [Africans] who were still living in the neighbourhood'.[23] The following morning a crowd of Africans were on site and they pointed out the graves of their people. Seventy-six old graves were not specifically claimed, so Hoffman proceeded to excavate all of these in the next two days. The skeletons were accessioned to the National Museum collection on his return to Bloemfontein. Hoffman was very pleased with himself because he felt that he had 'a good bunch of Bantu skulls now'.

Hoffman left the employ of the National Museum in 1946 after a disagreement with Van Hoepen, but returned to take over the directorship in 1951 when Van Hoepen was removed.[24] As director, Hoffman became obsessed with the search for pure representatives of the San and Khoe (Du Pisani 1989, 5). He reopened Dreyer's excavations at Matjes River and Florisbad and made research facilities available to Dreyer once again. Dreyer had been an honorary curator in the museum since the 1930s, but relations between Dreyer and Van Hoepen were so tense that Dreyer did not have free access to the collections of his own specimens (Van der Bank 2000, 116). Hoffman's obsession with the search for pure Khoekhoe led to one of the strangest expeditions sponsored by the museum. The Hoffman-Angola Expedition was the direct result of Hoffman following up on earlier research by Dreyer.

Early in 1935 the journal of Hendrik Jacob Wikar was published for the first time (Mossop 1935). Wikar had travelled along the

Orange River in 1778 and 1779 and this was his account of what he saw around the Augrabies Falls and a description of the people whom he met there. Of great importance was Wikar's observation that the people he met near the Falls were the first real 'Hottentots' he had encountered on his travels. Dreyer was among the readers of this new publication and Wikar's observation about real 'Hottentots' struck a chord with him. In June 1936 Dreyer, some African labourers from the museum, and his young colleague from Grey University College, Meiring, set out on an expedition to collect old 'Hottentot' skulls. Their purpose was to find the graves mentioned in Wikar's account (Hoffman 1972, 274) and their subsequent publication on the find included the first description of what was in their opinion the purest form of 'Hottentot' (Dreyer and Meiring 1937). Their expedition rapidly surveyed the banks of the lower Orange River from Augrabies Falls to beyond Upington and excavated any concentration of graves that they found. The largest cluster of graves was near the town of Kakamas and this was the name that Dreyer and Meiring would give their newly defined 'Hottentot' type. In the space of three weeks, they opened 112 graves and added 69 skeletons to the collection of the National Museum (Morris 1995, 111). Dreyer kept basic records, but these were not transferred to the museum along with the skeletons and it was only the fact that Meiring kept a private notebook with his own records that has enabled us to work out where the graves were disinterred and gives us a brief description of the style of grave.[25] When Dreyer and Meiring published their findings in 1937, the descriptions of the graves took on a special importance because they were used to classify the burials into 'pure' and 'degenerate' (Morris 1995, 111).

Meiring (1899–1986) was the only student that Dreyer taught who eventually followed him into the study of anthropology. Meiring obtained his Bachelor of Science degree in zoology and geology under Dreyer's tutelage and then went on to complete a Master's in 1924 in the same subjects. He briefly acted as a demonstrator of zoology

Figure 4.4: Illustration from Thomas Dreyer and Albert Meiring's 1937 article on the Kakamas burials from the Orange River. The pictures show two unexcavated graves, one skeleton exposed in the excavation of a grave, and a portrait of a Khoekhoe man who lived in the vicinity of the archaeological sites at the time of the excavations. (Source: Dreyer and Meiring 1937)

before breaking off his academic studies to farm near his hometown of Edenburg in the Orange Free State (Obituary 1986, 2). He returned to academic life as assistant to Dreyer in zoology at the recently renamed University College of the Orange Free State, where he completed a doctorate in 1936 under Dreyer's supervision. The subject of his doctorate was the Matjes River crania and he argued that these prehistoric South Africans represented a descendant branch of the Aurignacian people of Palaeolithic Europe (Meiring 1937, 69). He split his career between teaching posts at the University College of Fort Hare (1938–1951 and 1960–1968) and the museum service (assistant director at the National Museum in Bloemfontein between 1951 and 1957 and then director of the State Museum in Windhoek between 1957 and 1960). During his term as assistant director in Bloemfontein, Meiring collaborated closely with Hoffman and they reopened Dreyer's old workings at Matjes River and Florisbad. It was discussions about the definition of 'Hottentot' people with Meiring that triggered Hoffman's Angolan expedition.[26] Hoffman was determined to find the living descendants of the people that Dreyer and Meiring had found in the graves at Kakamas. Although the expedition was nominally under the auspices of the National Museum, Hoffman raised more than £3 000 from private and government sources, which made this a well-equipped expedition indeed (Hoffman 1960, 65). The trip was made by ten people in three Land Rovers travelling nearly 10 000 kilometres from Bloemfontein to Victoria Falls in what was then Rhodesia, across the Western Caprivi and into south-east Angola, returning via Windhoek and Keetmanshoop (Dreyer 1988, 9). Although Hoffman admitted that the expedition had failed in its primary goal of identifying the lost 'Hottentots' of Dreyer and Meiring, he found that the San people were located much further north than previously suspected. The expedition members expressed concern that the 'Bushmen' were hybridising with other native peoples and they predicted the 'total disappearance of these people in the foreseeable future'.

BROOM AND THE COLLECTION IN KIMBERLEY

By the 1930s, the museum in Bloemfontein had become known as a centre for archaeology, but some 165 kilometres to the north-west, the McGregor Museum in Kimberley did not have curators who actively researched the subject. Nonetheless, the human skeleton collection at the McGregor Museum grew to more than 130 human skeletons from the surrounding region before 1950 (Morris 1992a, 6). The collection at the museum indicates that the Riet River provided a particularly rich assembly of skeletons. Most of the Riet River graves were located and excavated by William Fowler between 1922 and 1946. Fowler was employed as the water engineer for the town of Koffiefontein in the Orange Free State, but also dabbled in local archaeology (Morris 1992a, 6; 1992b, 25). He did not keep careful notes of his discoveries, but there was generally a brief description in the museum records for each grave he excavated. Most of the excavations were close to Koffiefontein. Anthony Humphreys (1972) examined 79 Riet River graves and I examined 83 skeletons, the bulk of which came from the Fowler collection in Kimberley (see Morris 1984).

The Kimberley catalogue lists the brief data provided by Fowler, but the old catalogue has an additional assessment of racial type for many of them. This racial typology was the handiwork of Robert Broom (1866–1951), who for a time acted as the unofficial curator of the human skeleton collection in Kimberley. Broom had resigned from his academic post at Victoria College, Stellenbosch, in 1910. Findlay (1972, 29) suggests that the reason Broom gave up his academic career was because the pay at Victoria College was meagre and could not meet his financial needs at the time, and because Broom's scientific relations with the South African Museum had soured, especially because of disagreements with Péringuey. Broom briefly served as a general practitioner at Springs in the Transvaal and then returned to Britain via an extended stay in the United States in 1913. He did some further surgical training and then war service as a medical officer until 1916. When he returned to South Africa at the end of 1916, Broom was offered a temporary medical post in the town of Douglas in the northern Cape about 120 kilometres south-west of Kimberley. The temporary post was formalised as district surgeon in 1918 and he was resident in Douglas until 1929, the longest single placement Broom held in his entire medical career. Although his employment was in medicine, he continued to engage with scientific research, but his interests shifted from mammal-like reptiles to human craniology.[27] Not only did he visit the collection at the McGregor Museum in Kimberley, but he also actively collected skeletons for both the McGregor Museum and the American Museum of Natural History in New York. One of his last acts before leaving Stellenbosch had been to sell his collection of mammal-like reptile fossils to the New York museum and he continued this activity with human skeletons while in Douglas. Morongwa Mosothwane (2013, 29) identified 43 accessions linked to Broom at the American Museum of Natural History. Writing to the director of the museum in 1923, Broom said:

> Next week I am hoping to go off for a day on a Bushman hunt. I have just heard of a very promising Bushman burial place. It is near the spot where I got my two best Bushman skulls. I am sending you as a Christmas gift, one of these two Bushman skulls. It is the purest Bushman skull I have yet come across. With a little restoration it can be made into a fine exhibition specimen for your Hall of Primitive Man. Of course, I keep picking up all I can as in a few years Bushman skeletons will be unprocurable. The bones do not seem to last in our riverbank soils more than about 200 years.[28]

Broom developed a special interest in the Khoesan (Štrkalj 2000, 115). He published three important papers (Broom 1923, 1929, 1941) in which he not only elaborated detailed concepts of 'Bushman' and 'Hottentot' types, but also added a new type specimen for the Korana people, whom he thought were representatives of a very ancient physical type in South Africa who had mixed with the San to form the Khoekhoe. In Broom's mind the Korana displayed very robust physical features, including a strongly developed region about the eyes (supra-orbital torus), which is a common feature in Australian Aboriginals. Broom hypothesised an Australoid prehistoric group who were no longer present, but had lived on several continents before becoming extinct everywhere except in the Australian region. The recently discovered Rhodesian skull (recovered from Kabwe in Zambia in 1921) was an example of these ancient Australoids, and the robust features of the Korana were evidence to Broom that they had mixed with the 'Hottentots' of old (Broom 1923, 145).

Broom applied the same typological methods he used in his studies of mammal-like reptiles to his studies of humans. He searched for racial essences in the visible morphology of the crania of his archaeological skeletons to reconstruct past racial history from these features. For example, in his 1941 paper, he reported on 32 of the Riet River skulls, of which he classified 15 as pure 'Korana', 9 as pure 'Bush' and 8 as 'Korana-Bush' hybrids (Broom 1941, 239). The entries in the Kimberley catalogue suggest that he went even further by identifying other individuals as being from ethnically specific living groups even though all the skeletons were from prehistoric archaeological contexts.

TYPOLOGY AND NATIONALISM: WAS THERE AN AFRIKANER PHYSICAL ANTHROPOLOGY?

The central problem of typology as applied to South African populations was that it was entirely subjective and there was no statistical methodology to verify the observations. Each researcher simply decided on which characteristics were diagnostic of type, race or culture. Dreyer (1931) tried to engage with this problem by surveying the literature available to him at the time to try to find acceptable research questions about the origin and differentiation of Khoesan peoples. When Leonard Gill, director of the South African Museum, sent a prepublication copy of Dreyer's article to Dorothea Bleek, he noted: 'It states the problems quite well, but when it comes to the author's own contributions it becomes so incoherent that it is very difficult to follow.'[29] When the paper was finally published, Broom made an angry note in the margins of the McGregor Museum library copy, complaining that Dreyer had not properly reported on Broom's data.[30] When Dreyer and Meiring published their Kakamas descriptions (1937), they chose a set of five skulls as the Kakamas type, to represent the descendants of Hamitic migrants from the north who formed the core of the 'Hottentots'. They based their choice on three features: (1) the map location of the graves where Wikar had met the real 'Hottentots'; (2) the fact that these five skeletons were found under high complex cairns that looked like North African inhumations; and (3) the long and narrow morphology of the skulls, which was very unlike that of 'Bushmen' (Morris 1995, 111). Inherent in their discussion was their idea that the high cairns were signs of pure culture and that lower cairns, or less formalised graves, were signs of cultural degeneration. Cultural behaviours were seen as a mark of underlying racial characteristics.

The link between behaviour and biology was a keystone in arguments about race at this time. Heinrich Vedder, the German missionary ethnologist, provided a telling description of this link in his chapter on the Khoekhoe in his book *South West Africa in Early Times*:

> There were, however, always two kinds of blood in their veins, the Hamitic blood, in which lay the urge to stock-raising and the love of animals, and the Bushman blood, to which stock-raising was something entirely foreign but, for this very reason, made them all the keener upon hunting. This is why, in the case where the Bushman blood persisted, there is still to-day an aversion to stock-raising and, accompanying it, a propensity to despoil the flocks of others and to treat them as wild game. (Vedder 1938, 123)

When some of Dreyer's and Meiring's ideas about the Kakamas type were debated by Broom and Lawrence Wells, they accused these foreign-born researchers of not understanding the term 'Hottentot' in the same manner as 'used by South Africans born in the country'

(Dreyer and Meiring 1952, 19). Dreyer took the criticism personally and when towards the end of his life he was offered an honorary degree from the University of the Witwatersrand (Wits), he initially refused because he did not want to be linked to Dart's school in Johannesburg.[31]

Dreyer's prickly response was a sign that the political question of Afrikaner nationalism had become an issue in Bloemfontein. Dreyer did not take a specifically political view. His daughter, Marie Barry, remarked that her father was a supporter of both the National Party and the South African Party in the 1930s and saw himself as a South African patriot, not a political patriot.[32] He privately thought that the implementation of apartheid was unnecessary because it already effectively existed. Dreyer saw the advancement of South African science as the issue, not politics.

Dreyer clashed with his colleagues at both Grey University College and the National Museum over the use of Afrikaans in teaching and publication. Dreyer had already lost an opportunity to be rector of the college because of his preference for writing and teaching in English, and this was to become worse in the 1930s and 1940s as the college moved towards its new status as a fully-fledged university. Dreyer was a vocal supporter of the group that unsuccessfully petitioned the government to make it a home for both English and Afrikaans speakers (Morris 2012, S157). His views on Afrikaans were not shared by Van Hoepen as director of the National Museum. Van Hoepen was an extremely difficult man, who often fought with his staff and the museum council, but he was also adamant that the museum and its researchers should publish in Afrikaans and not in English. Van Hoepen had been in active competition with Goodwin and Van Riet Lowe in the late 1920s to develop a South African nomenclature for archaeological cultures and technology. Van Hoepen wanted the new terminology to be in Afrikaans (Underhill 2011, 5). He asked Meiring to submit his thesis for publication in Afrikaans for the *Argeologiese*

Navorsing van die Nasionale Museum. Meiring made several attempts to convince Van Hoepen to accept the article in English and in the end did not submit in Afrikaans. Van Hoepen was furious and he complained while on a visit to Johannesburg that Meiring came to the museum too often and that he must not have enough to do at the college. Word of Van Hoepen's complaint eventually got back to Dreyer, Meiring's boss, and Meiring found himself writing to Alexander Galloway as a third party to inform Dreyer that this was just Van Hoepen grumbling. He noted in his letter to Galloway that he 'didn't mention Dreyer's name to van Hoepen because you know they're not exactly pally!'[33]

Perhaps the best way to describe Dreyer would be to include him as part of the movement toward the 'South Africanisation' of science (Dubow 2000, 66; 2006). Although these researchers perceived science as being within the domain of white South Africans, their main desire was to put South African science and scientists on the world stage as equal to the science of their European mentors (Dubow 2006). Many, like Van Hoepen, held explicitly racist views, such as his support for white rather than black labour when he ran for a seat on the Bloemfontein city council in 1931 (Van der Bank 1998, 107), but their nationalism went beyond racial determinism. Dreyer wanted to use science to improve the lives of South Africans. He was a convert to the grand plan of Ernest Schwarz to redirect the waters of southern African rivers to flood the Kalahari, so that rainfall would be increased and agriculture throughout the subcontinent improved (Eales 2007, 130). Schwarz had originally proposed his ideas in a 1919 book, but also gave a public presentation at the Bloemfontein S2A3 conference in 1923, for which Dreyer had been on the organising committee (Schwarz 1923). When a group of farmers and academics created the Kalahari Thirstland Redemption Society during the drought of 1933, Dreyer joined them as treasurer. Even though three government investigations over 30 years had declared Schwarz's ideas impractical, farmers, business people and others continued to support the proposal through the Society and by means of letters to newspapers and to the journal *Farmer's Weekly*. Meredith McKittrick (2015, 501) has examined why Schwarz's ideas continued to attract followers even though he had died in 1928. In McKittrick's view, the Kalahari Thirstland area was part of a greater South Africa, made up of South Africa and the neighbouring countries whose lands would provide water supplies and the agricultural future of South Africa. Although a formal claim to territory outside of South Africa's borders (except for Namibia) was not under consideration, the supporters of the Redemption Society saw the resources of neighbouring countries as part of the regional homogeny of South Africa itself. This was science in the service of white South Africa.

Language and geography were certainly concerns of the anthropological scientists in Bloemfontein, but a publication by Dreyer, Meiring and Hoffman in the German scientific journal *Zeitschrift für Rassenkunde* in 1938 suggests that there may have been darker anthropological theories at stake. The 1938 article was different from anything the three authors had published previously. In it, they presented a Mendelian genetic argument for the inheritance of racial characteristics and the formation of mixed racial groups. To understand how this particular piece of writing came about, we need to look back not at the training and research focus of Dreyer or his student Meiring, but to the Stellenbosch education of the museum scientist Hoffman.

After the resignation of Broom in 1910, there was little teaching of anthropological subjects at Victoria College in Stellenbosch. Grobbelaar, who had been one of Broom's students at the time of the British Association Meeting in 1905, and had remarked on how Broom's lectures inspired him in anthropology (Grobbelaar 1955, 324), returned to the college as a lecturer in 1913 and completed his Master's in zoology while teaching. He almost certainly would have taught Brink his zoology as an undergraduate in 1916, but Grobbelaar did not have a higher degree allowing him to be promoted to professor. He took leave in 1922 and 1923 to complete a doctorate in Berlin. Cornelis Gerhardus Stephanus (Con) de Villiers was another Master's student at Victoria College at the same time as Grobbelaar. After completing his Master's in zoology, De Villiers taught at the Transvaal University College, but left for Zurich in 1918 to do his doctorate. His subject of choice was zoology, but while at Zurich he studied anthropology under Otto von Schlaginhaufen (Walters 2018, 57). Schlaginhaufen himself had been a student of the famous physical anthropologist Rudolf Martin and had become professor at the University of Zurich, taking over from his mentor Martin in 1911 (Ziegelmayer 1997, 912). De Villiers returned in 1923 to the University of Stellenbosch (now constituted as a full university from its forerunner Victoria College) to take the post of professor of zoology, the same year that Grobbelaar returned from Berlin with his doctorate, and the same year that Dart arrived in Johannesburg to take up the post of professor of anatomy at Wits.

The appointment of De Villiers as professor at Stellenbosch marked some significant changes in the course offerings in zoology. Grobbelaar's post was expanded to senior lecturer in zoology and physical anthropology, but it was De Villiers's experience with Martin's methodology in Zurich that was informing the course. In 1924 the university launched a new course in anthropology taught by De Villiers and Grobbelaar. The new course used three primary textbooks: Martin's *Lehrbuch der Antropologie*, published in 1914; Harris Wilder's 1921 *Laboratory Manual of Anthropometry*; and a 1923 English translation of Marcellin Boule's *Les hommes fossiles* (Walters 2018, 25). The course had a specific focus on practical anthropometry. The department purchased Rudolf Martin's eye, Eugen Fischer's hair and Felix von Luschan's skin colour standardisation charts and began to study the physical variation of

Stellenbosch students. Handri Walters has looked in detail at these courses and the publications that resulted from their research. In her opinion, this research intended to infer 'a biological superiority of the white Afrikaner race by likening it to the superior races of Europe' (Walters 2018, 64). The same books and anthropology charts could be found in Drennan's and Dart's departments (at UCT and Wits, respectively), but there was not the same focus on measuring large numbers of white students. Dart and Drennan looked at South African prehistory in their science courses, and projects were more often linked to anatomical variation, rather than to racial features (although some racial studies were done).[34] Perhaps the strongest difference between Stellenbosch and the other schools is that physical anthropology did not spread as a topic into the associated ethnology and social anthropology departments in Johannesburg and Cape Town. Stellenbosch offered a course called 'Introduction to Human Heredity' to students in the Department of Ethnology between 1927 and 1930 and a third-year practical course in physical anthropology, racial hygiene and native administration between 1931 and 1936. Grobbelaar was so involved in physical anthropology that he planned to go to Munich to study anthropology in 1939. When his plans fell through because of the war, he spent a year at UCT studying physical anthropology under Drennan (De Kok 1987, 307). Grobbelaar taught physical anthropology to the students in the new Department of Physical Education when it was launched in 1942, and the students were expected to be able to read the original German in the classical and modern literature of the subject (Walters 2018, 69). By the end of the 1930s, the research focus of the department had moved from assessing variation of white university students to a broader project on the biological definition of the coloured population. Walters (2018, 79) has looked at the motivation for these kinds of studies at Stellenbosch, and for her, the research moved from the study of purity (the comparison of South African whites to Europeans) to the study of racial mixing (coloured people as a product of miscegenation).

The coloured population was an ideal laboratory for exploration of the question of racial mixing, but not until substantial research had been done to describe the full range of variation in coloured individuals. The chief mechanism for the analysis would be the new German approach to racial biology outlined by Fischer in an exhaustive study of the 'Rehoboth Basters' (1913) as an example of a Mendelian first-generation cross between two races. The approach fused Mendelian genetics and anthropometry and was based on a theory of particulate inheritance, according to which racial essences could be passed on. Physical or mental traits, no matter how complex, were interpreted as unified entities whose inheritance was unaffected by environmental conditions and determined by single genes. Fischer believed that racial traits could be dominant and recessive and that dominant characters were more strongly expressed in mixed-race populations (Proctor 1988, 146). He also believed that behavioural traits were also governed by Mendelian inheritance, so racial behaviours could be inherited in the same manner as physical characteristics. Racial genetics assumed that the simple inheritance demonstrated in Gregor Mendel's pea plants was true for all or most traits. Mendelian analysis demanded that the parental populations were homogeneous, something that is just not present in humans at any level, but even more so because the racial traits under study were all multigenic with extremely complex inheritance (Hildebrandt 2009, 899). The fundamental flaws in racial genetics were overlooked and the method produced the results that the race scientists wished to see.

When Brink returned to South Africa to practise medicine in the late 1920s, it was to De Villiers at Stellenbosch, rather than Drennan in Cape Town, to whom he turned for supervision on his renewed Griqua project. For De Villiers, this was a great opportunity to look at a first-generation cross and put Fischer's Mendelian genetics to the test. Brink was fully aware of Fischer (1913) as the work is referenced in his 1923 paper (written while he was still in England), but only after his contact with De Villiers did he try the Mendelian genetic approach to analyse race crossing (Brink 1933, 14). The methodology used by Brink in 1933 is identical to the methodology used by Dreyer, Meiring and Hoffman five years later. De Villiers's methodological influence cannot be disputed, but it remains uncertain how extensively his broader racist vision was passed on to his students. De Villiers travelled to Germany in the 1930s to give lectures and meet with German academics in Göttingen, Leipzig, Munich and Breslau. On his return, he not only brought the ideas of German anthropology to South Africa but also German racial nationalism. De Villiers was one of the protestors who objected to the arrival of Jewish refugees aboard the SS *Stuttgart* in October 1936 (Walters 2018, 57). The delegation of five University of Stellenbosch colleagues, led by Professor Hendrik Verwoerd, made a formal protest to the United Party government and demanded laws to prevent Jewish immigration to South Africa (Robins 2016, 32).

What is intriguing is that the German approach to racial genetics did not pass into the English-language schools in Cape Town and Johannesburg. There were certainly publications on racial genetics in English, especially from the geneticist Reginald Ruggles Gates (1926, 1929), but in general the anthropologists who were creating the typologies of modern and prehistoric peoples were more interested in descriptions of the crania than the mechanism of inheritance. Whereas the physical anthropology in the English-language schools meshed closely with anatomy, physical anthropology in Stellenbosch was closely linked to the racial politics of *volkekunde* (folk studies, ethnology) and eugenics (Walters 2018, 97). Dart, Drennan, Wells, Broom, Dreyer and Galloway were not members of the Eugenics Society of South Africa. For the eugenicists, the study of race was really about the white race and how racial hierarchies, with whites at the top, could be maintained (Dubow 2010, 282). For the prehistorians, the study of race history was about finding the ancient pure races. The products of miscegenation, especially in modern populations, were simply complications that made the identification of racial types more difficult.

The data in Dreyer, Meiring and Hoffman's 1938 paper were drawn from the work of Dreyer and Meiring, but it was Hoffman who provided the analytical technique. Whereas Dreyer and Meiring continued to focus on their Kakamas type and the origin of the Khoekhoe, Hoffman became more and more intrigued with the questions being explored by De Villiers during his time in Stellenbosch. Hence Hoffman's expedition to Philippolis to collect more Griqua skeletons and later in his expensive and very public expedition to Angola to find evidence of the last of what he considered to be pure 'Hottentots'. Hoffman was very disappointed and pessimistic when he did not find pure populations, but instead witnessed the 'degeneration' caused by the mixing of races (Du Pisani 1989, 5). Like his Stellenbosch mentor, Hoffman was concerned about the negative outcome of sexual union between different groups of people, although he did not formally link this to the race policy of the government that was in power at the time.

The typological recipe that researchers applied to the peoples of southern Africa was drawn directly from the European anthropological cookbook. The ingredients were the range of types created to describe the complex patterns of economies, languages and physical variations of both early historic and modern populations. Behaviour and biology were mixed. Whether it was mysterious Hamites who had migrated down from the north as the 'Hottentots' or the big-brained Boskop people who were thought to be the Cro-Magnons of Africa, each people came with identifiable physical features and a culture to match. Where a group was not clearly identifiable as one type or another, its ancestry was assessed as being made up of strains left over from ancient types now extinct, such as Broom's Australoids. The typologists were so intent on finding these strains that they sometimes massaged their data to enhance them. Drennan would ignore and exclude the skeletons from Colesberg that did not fit his typological vision of 'Bushmen', while Galloway defined 'Negroid' characteristics in such a way that he managed not to see any of them in his study of more than 30 skeletons from the Iron Age site of Bambandyanalo. The legacy of typology was this fixation with racial purity and its identification. These researchers were trying to 'unscramble the egg' to reveal its underlying constituents to help them understand the bewildering complexity of human variation.

NOTES

1   Interview with Marie Barry (née Dreyer), 29 May 1993, Oudtshoorn.
2   Interview with Baz Edmeades (Dreyer's grandson), December 1992, Cape Town.
3   Interview, Edmeades.
4   Interview, Barry.
5   National Museum, Bloemfontein, Thomas Dreyer, 'Introduction', 1981.
6   Iziko Museum, Cape Town, letter from Louis Péringuey to Thomas Dreyer, 17 August 1915.
7   Interview, Barry.
8   'Dr Dreyer Unearths Some Prehistoric Relics', *The Friend*, 27 May 1930.
9   Albany Museum, Grahamstown, letter from Thomas Dreyer to John Hewitt, 28 September 1930.
10   Department of Human Biology, University of Cape Town, letter from Thomas Dreyer to Matthew Drennan, 30 October 1930.
11   'Local Collection of Fossils', *The Friend*, 14 November 1931.
12   *The Friend*, 13 August 1932; *The Star*, 10 March 1933.
13   National Museum, Bloemfontein, letter from A.J.D. Meiring to J.J. Oberholzer, 20 April 1981.
14   National Museum, Bloemfontein, 'Reports on Monograph'.
15   David Morris, McGregor Museum, letter to author, 17 January 2020.
16   Handri Walters and Karlien Breedt, personal communication, 6 December 2019.
17   Copy of letter from Vernon Brink to Howard Phillips, 29 March 1983, in private possession of Professor Howard Phillips, Cape Town.
18   Natural History Museum, London, letter from Vernon Brink to Arthur Thomson, 13 September 1923.
19   Iziko Museum, Cape Town, letter from Vernon Brink to Leonard Gill, 27 June 1939.
20   Department of Human Biology, University of Cape Town, note from James Drury to David Slome, undated.

21 National Museum Library, Bloemfontein, *Ekspedisieverslae*, 'Ekspediesie-rapport #60', 12 November 1934.

22 National Museum Library, Bloemfontein, *Ekspedisieverslae*, 'Ekspediesie-rapport #108', 25 September 1941.

23 National Museum Library, Bloemfontein, *Ekspedisieverslae*, 'Ekspediesie-rapport #115', 15 April 1942.

24 Van Hoepen was constantly at odds with his museum council and there were several calls for his removal as director. He was finally dismissed from his post due to financial irregularities and his failure to adequately make the museum an educational (rather than research) institution (Badenhorst 1968, 700).

25 National Museum, Bloemfontein, A.J.D. Meiring, 'Boek met Veld nommers van grafte, Kakamas & Augrabies [Book with field numbers of graves, Kakamas & Augrabies]', 1936, handwritten note on a University College of the Orange Free State examination booklet.

26 As late as 1977 Meiring still believed that the Kakamas graves represented 'an interesting race that once occupied large portions of southern Africa … from SWA [South West Africa] down and around the southerly tip and up in the eastern side perhaps as far as the present Port St Johns'. Department of Human Biology, University of Cape Town, letter from Albert Meiring to Alan Morris, 24 August 1977.

27 Department of Human Biology, University of Cape Town, L.H. Wells, 'Robert Broom: From Country Doctor to World-Famous Palaeontologist', presented to Medical History Club, 17 September 1973.

28 American Museum of Natural History Archives, New York, letter from Robert Broom to Henry Fairfield Osborn, 23 November 1923.

29 Iziko Museum, Cape Town, letter from Leonard Gill to Dorothea Bleek, 16 July 1931.

30 On page 82 of the McGregor Museum's copy of Dreyer (1931), Broom wrote: 'There is not a word of truth in this. The Korana was prisoner in Douglas gaol and died there. He was buried by my own hand in my garden. So surely I know all about him. I have now in my possession his whole skull and skeleton. The skull I dug up on "Boy Cilliers" farm was that of a Morolong who was drowned in crossing the Orange River. I don't have his skull but it has now been figured. Further comment is unnecessary. R. Broom.'

31 Interview, Barry.

32 Interview, Barry. Dreyer's grandson was a bit stronger on this point, stating that Dreyer 'had one ball in the Nats and the other in the SAP'. Interview, Edmeades.

33 University of the Witwatersrand Archives, Johannesburg, letter from Bertie Meiring to Alexander Galloway, 16 April 1937.

34 A good example is S. Matus's 1941 project on the 'Mongol Spot', supervised by Drennan and looking at the frequency of lower-back skin discolouration in African and coloured populations in Cape Town.

5

Raymond Dart's Complicated Legacy

Raymond Dart did not want the job as head of the Department of Anatomy in Johannesburg. Aged 29 in 1922, Dart was enjoying the patronage of Grafton Elliot Smith in his laboratory at University College, London. After graduating from university in Sydney, Dart had joined the Australian Imperial Forces. He arrived in Europe too late to be involved in active warfare, but his medical skills were in demand as the influenza epidemic began to take hold among the soldiers. As his period of military service neared its end, Dart was offered the position of demonstrator in Smith's Department of Anatomy through the influence of James Wilson, his old professor from Sydney. Although it was rarely given, Dart requested and received permission from the Australian army for demobilisation without returning to Australia and in October 1919 he joined Smith at University College (Wheelhouse and Smithford 2001, 40).

Under Smith's guidance, Dart was introduced to a range of new research topics, including neuroanatomy and anthropology. Smith also recommended Dart, along with another young Australian anatomist, Joseph Shellshear, for the new Rockefeller Fellowship in the United States. The two spent six months teaching anatomy in American departments, three months visiting other schools and three months at the Woods Hole Research Institute in Massachusetts. The trip was a wild success and Dart was especially impressed with Robert J. Terry's collection of skeletons at Washington University in St. Louis (Tobias 1984, 7). Upon their return in late 1921, Shellshear took up the chair of anatomy in Hong Kong, but Dart continued in Smith's laboratory in London.

Although he had yet to make a firm decision about his future, Dart was enjoying the research in London, especially working on the neuroanatomical material that was Smith's speciality. He assumed that he would stay on in London and he could hardly have been more disappointed when Smith proposed that he take up the post at the University of the Witwatersrand (Wits). Dart feared that in far-away South Africa he would be too far from the centre of scientific research that England represented (Dart and Craig 1959, 31). London was the venue for the English meetings of the Anatomical Society of Great Britain and Ireland, the Zoological Society of London and the Royal Society of London and this is where he could network with the finest minds in anatomical science. There would be nothing like that in Johannesburg. Dart turned down the proposal and it took substantial persuasion from Smith to get him to change his mind. Dart was interviewed on behalf of Wits by Smith, Arthur Keith and James Wilson. All three spoke of the glowing opportunities in South Africa (Wheelhouse and Smithford 2001, 50). The university senate in Johannesburg accepted his application, but only as a second choice after the preferred candidate, Robert Thomson (the previous head of anatomy in Cape Town), had withdrawn (Tobias 1990, 334).

Dart was right to be apprehensive. His fears were justified when he found out how few resources were available to him when he arrived in

Johannesburg. Much of the city still had the appearance of a mining camp – its buildings with their red, corrugated-iron roofs seemed only temporary. The Medical School was situated in a two-storey building behind the walls of the old Johannesburg Fort and the grounds were untended and full of weeds. The dissection rooms were even worse, desperately needing paint and without running water, electric plugs, gas or compressed air for the student laboratories (Dart and Craig 1959, 33). The cadavers that formed the basis of the dissection teaching programme were badly stored and had become dried out. Some of the material was so hard and desiccated that it was impossible to dissect the ligaments of the human foot.[1] Dart's wife purchased new cotton cloth to properly wrap the bodies and Dart arranged for rubberised sheeting to shroud each cadaver (Dart and Craig 1959, 33). Even worse was that he did not receive a warm welcome from either the Medical School staff or students. Dart had been hired to replace Professor Edward Stibbe, who had been forced to resign by the principal, Jan Hendrik Hofmeyr, and the university council over his extramarital indiscretion (Murray 1982, 81; Tobias 1990, 427). Stibbe had been a popular lecturer and students perceived his removal as unfair. The university senate had disagreed with Hofmeyr, demanding that Stibbe's services be retained, but Stibbe had already resigned. The conflict between senate and council over the Stibbe affair resulted in the university adopting standard procedures for dismissal and suspension in which the principal's power was substantially limited (Murray 1982, 91). All of this was too late for Stibbe and the ill-feeling was still present when Dart arrived. In addition, Hofmeyr was prejudiced against Australians (Dart and Craig 1959, 34).

The only piece of good news was that Wits provided an additional post for a professional assistant, which was filled by Dr Gordon Laing who arrived from Aberdeen during Dart's first year. Dart was also lucky that the teaching crush of returned soldiers from the First World War was over by 1923 and the classes were of a more manageable size. A comparison with Matthew Drennan in Cape Town is enlightening, as Drennan faced classes of more than 300 students, with only transient student help in demonstrations. Stibbe had begun the teaching programme at the beginning of 1920 and, with student numbers increasing, the first MBChB (Bachelor of Medicine, Bachelor of Surgery) graduates would obtain their degrees in 1925 (Keene 2013, v).

Despite these first setbacks, Dart hit the ground running during 1923. The Johannesburg medical anatomy programme was spread over two years. Dart immediately shortened the medical teaching to one year of full dissection for both medical and science students (Anatomy I), and a second year (Anatomy II) for students interested in the exploration of scientific topics.[2] The Anatomy II course was very popular because of Dart's personality and boundless enthusiasm for research. Within a year, there was already a full load of students in Anatomy II. Having a full-time academic assistant helped a great deal, not only in spreading the teaching load, but also because Dart was able to interest his senior staff in research. When Frederick FitzSimons in Port Elizabeth brought the skeletons from the Tsitsikamma caves to Dart for study, Dart involved Laing in the analysis. Although Laing left the department some seven years later to become the medical officer of health for the City of Johannesburg (Tobias 1984, 10), he retained his interest in anthropology sparked by Dart and eventually registered at St Andrews University in Scotland for a doctorate in dental anthropology.[3] Laing's tenure was followed by Lewis R. Shore, who was a friend of Dart's from his London days (Tobias 1984, 10). Shore did not work on archaeological material, but he did do the first work on the growing collection of vertebrae from the cadaver skeletons in 1930 and 1931. Shore would be followed by Alexander Galloway, who arrived from Aberdeen via Canada in 1932.

There was so much public interest in the Tsitsikamma skeletons that Dart thought a further excavation done by him, rather than

FitzSimons, might be possible. For Dart this was about acquiring skeletal specimens for research, not archaeology, a field in which he had no experience. Early in 1927 he proposed to Wits's social anthropologist Winifred Hoernlé that they join forces to excavate an archaeological site as a student project. Hoernlé broached the idea with Professor Tom Barnard at the school of African Life and Languages in Cape Town, who thought the idea was wonderful.[4] Barnard suggested that he and John Goodwin could bring along two or three students and that he might even have a little money to cover some expenses. It was left to Dart to come up with an appropriate project.[5]

In October 1927 Dart approached John Phillips of the Deepwalls Research Station near Knysna about excavating caves in the area. Phillips responded with a letter providing details not only of possible cave locations, but also details of logistics for food, accommodation and local guides and labour.[6] Dart then wrote to Barnard in Cape Town with a formal proposal to excavate a cave at Krantshoek in an area some distance from the caves already sampled by FitzSimons of the Port Elizabeth Museum.[7] Of great importance was that Dart would obtain the services of James Drury of the South African Museum to direct the excavation, making the project a joint effort of three institutions: the University of Cape Town (UCT), Wits and the South African Museum. All that would be required was funding to support the project, for which Dart approached the university's Bantu Research Grant Committee for their blessing and a grant of £80 to £100 to make the expedition happen. Sadly, the committee did not agree. Not only did they refuse funding, but they specifically noted that they were 'sceptical of the practicability' of the project. In the opinion of the committee, the excavation should be under the auspices of only one institution.[8]

Dart's enthusiasm not only affected the staff, but also had a major impact on the students. The students registered for Anatomy II had already demonstrated academic expertise or they would not have been allowed to progress into the science programme. Many of the students began projects involving the new Tsitsikamma Strandlopers, as well as the growing collection of cadaver skeletons. Josephine Salmons would be among the first to describe Khoesan skeletons in the collection (Salmons 1925) and Harry Sutherland Gear would cut his anthropological teeth on the Boskop skeletons provided by FitzSimons (Gear 1925, 1926; Laing and Gear 1929).[9] There were many other outstanding students, but few went on to become anthropological anatomists. The most significant exceptions were Lawrence Wells (Bachelor of Science, 1928), Joseph Weiner (Bachelor of Science, 1934) and, of course, Phillip Tobias (Bachelor of Science, 1945), who would all go on to play a major role in the development of physical anthropology in South Africa.

Stibbe had not started an anatomy museum and the only demonstration specimens available to Dart when he arrived were fifteen disarticulated skeletons in cardboard boxes scattered around the department.[10] Dart initiated a programme on several levels to begin a substantial collection of both teaching and research specimens. In discussion with Drennan in Cape Town, Dart set out a new method of dissection in which all the bones of the skeleton were left intact and undamaged.[11] The only exception was the skull, where the braincase was cut transversely for the removal of the brain, but the separated calotte was kept and replaced after maceration. After the conclusion of the dissection programme, each complete skeleton was kept, along with the mortuary data from the cadaver records, in the manner of Terry's collection in St. Louis.[12] At the same time, Dart used his network of anatomical contacts to request comparative human skeletons from medical practitioners on the Witwatersrand. Dr Louis Fourie of the South West Africa administration was asked for 'Bushman' material and responded with the donation of at least one skeleton.[13] The publicity from Dart's newspaper publications about the FitzSimons discoveries resulted in the donation of chance archaeological skeletal discoveries to the collection. Dart had also recognised that baboon biology might be an interesting avenue of research and in 1924–1925 established a baboon colony on the roof of the Medical School (Keene 2013, 4).[14] Dart expressed the general consensus at the time that anthropological anatomy had to be rooted in comparative primate anatomy and his baboon colony would not only allow for the study of living monkeys, but also provide bodies of dead ones for detailed study. Just as important were the specimens brought in by his highly motivated science students. Dart had initiated a £5 prize for the student who brought in the best biological specimen as an exhibit. Dart's little competition to find specimens for the anatomy museum would ultimately provide Dart and Wits with anthropological fame.

The discovery of the Taung Child and the role of Josephine Salmons and the science class have been well described (Dart and Craig 1959, 1; Tobias 1984, 17; Wheelhouse and Smithford 2001, 59; Štrkalj 2005, 97; 2006a, 255). The events themselves are not important to this story, but the impact of the discovery on both Dart and the university is central. Dart rushed his now famous paper off to *Nature* (Dart 1925a) and the storm of publicity began as soon as the published journal became available. Dart was greatly disappointed that the great minds of British anatomy refused to accept the new specimen from Taung as a human ancestor (Tobias 1984, 38). In comments published in *Nature*, Keith relegated the specimen to the same family or subgroup as the chimpanzee and gorilla. Dart's mentor, Smith, hedged his bets until the geological age was identified and the anatomy of the teeth described. Arthur Smith Woodward, who along with Keith was a great supporter of Piltdown, felt the skull had little bearing on the ancestry of human beings. Wynfrid Duckworth at Cambridge was the most supportive, but even he reserved judgement until an adult specimen could be found some time in the future. Dart could not make headway in the acceptance of Taung without the support of these great

Figure 5.1: Raymond Dart in 1925 with the Taung skull shortly after its discovery. Dart was 32 years old and had been in the Department of Anatomy at the University of the Witwatersrand for just over two years. (Source: Tobias 1984)

men of science and he realised that he would have to travel to England with the specimen to argue his case.

Dart himself described some of the public commentary (Dart and Craig 1959, 43) and his disappointment in the scientific rejection of his primary conclusions, but he also noted how the South African scientific community, and especially his colleagues at Wits, recognised the importance of the discovery and accepted him as an important scientist. Within a year he was invited to become president of the anthropological section of the South African Association for the Advancement of Science (S2A3) and a Fellow of the Royal Society of South Africa. Further recognition came when he was elected dean of

the Faculty of Medicine. In fact, it was Dart's discovery and his vig-
orous defence of its interpretation that brought international fame, if
not support. Bruce Murray goes further and says that Dart 'put the
medical school, and indeed the University truly on the map' (Murray
1982, 179).

The years after the discovery of the Taung Child were busy ones for
Dart. As dean of the Faculty of Medicine, he was heavily involved in
the development of the Medical School, especially its library. He was
also a member of the Loram Committee, which was examining the
possibility of training black medical doctors (Murray 1982, 303) and
he continued to teach both medical anatomy and science students.
Despite this heavy load, Dart continued his research, publishing 24
papers between 1925 and the start of his sabbatical in 1929 (nearly
5 papers per year).[15] He gave the presidential address of Section E
(Anthropology) at the S2A3 meetings at Oudtshoorn in July 1925.
At this time Dart had been in Africa just two and half years, but his
address had an Afrocentric focus, which was quite different from his
contemporaries, although still clearly within a white South African
view (Dart 1925b, 75). He reported on recent discoveries, including
Boskop and his own work on the Tsitsikamma skeletons and Taung,
and he talked about the great time depth of humanity in Africa – at
least equal to that of Europe.

He then focused on the study of race, specifically of the 'Bushman'
and 'Bantu', which he saw in comparative terms with Europeans.
He felt that this would lead to discoveries in surgery and medicine. He went on to highlight the question of the great mysteries of
African culture, specifically the riddle of the ruins of Zimbabwe, the
terraced cultivation of Nyanga and ancient metal mining, which in
his opinion demonstrated Africa's contacts with the empires of the
old world. These two topics, race and cultural diffusion, would consume most of his research focus in the coming years. Dart lamented
the European focus of South African universities (which in 1925

would have included Greek and Roman Classic studies, European politics and literature) and made a plea for local universities to begin to build an 'African philology, African ethnology, African archaeology and African anthropology' (Dart 1925b, 80). He may have been influenced in this view by liberal radicals at Wits, such as William Miller Macmillan and Margaret Hodgson (history) and Winifred Hoernlé (Bantu studies) (Murray 1982, 125). Dart also noted: 'There has grown up over the recent historical period amongst the Nordic peoples a bitter intolerance of those possessing skin colouration. This phenomenon, which is based on political sentiment and has no justification in biological laws ... is nevertheless very real. What is more, it is destined to be the most pressing problem in the future of humanity' (Dart 1925b, 79). This appears to be Dart's first and last venture into the politics of race.

Dart's relations with the S2A3 were extremely important in these early years and it is clear how much he saw this organisation as the spearhead of South African science. Dart had joined the S2A3 in the year of his arrival and was to stay on as an ordinary member of council from the time of joining until his war duties in the Field Ambulance Corps took up what was left of his free time in 1940. The S2A3 meeting for 1929 was scheduled to be held in Johannesburg as a special joint gathering of the South African and British associations. The last joint meeting, held in 1905, had been an opportunity for South African science to benefit from the guidance of the older institution (Morris 2002, 337), but the South African delegates for the 1929 conference were determined to show how far South African science had advanced in the intervening 24 years (Dubow 2000, 66; 2006). Hofmeyr's presidential speech addressed this issue, identifying physical anthropology specifically (Dubow 1995, 52). Dart and his students made a special effort, with no fewer than five papers by four of Dart's students and three of his own. Two of Dart's papers require special mention here. His paper on the Taung skull (Dart

1929a) speculated about the geological age, which had been a sticking point identified by international reviewers, while the second (Dart 1929b) highlighted a special interest of his – the presence of phallic symbolism in discoveries in Zimbabwe and associated sites. Dart had special reason to talk about Zimbabwe because the conference also hosted Gertrude Caton-Thompson, who had just returned from excavating at the ruins of Great Zimbabwe and was about to report on her finding that they represented the material remains of an entirely local African civilisation not more than 400 or 500 years old (Caton-Thompson 1983, 128).

Caton-Thompson's presentation attracted wide public as well as scientific interest and her audience on 2 August 1929 had several foreign delegates, including Leo Frobenius of Germany, as well as a contingent of newspaper reporters. Various views were expressed when the floor was opened to questions at the end of her paper, some supportive and some not. Frobenius argued for his Erythrian theory, in which Zimbabwe would be linked to an ancient Middle East influence at least 4 000 years older than Caton-Thompson's model. However, it was Dart who caused the greatest upset at question time. Ignoring the Frobenius input, 'Professor Dart, in a fierce outburst of curiously unscientific indignation with the whole course of the discussion, charged the chairman with having called upon none but supporters of the Caton-Thompson theory to speak'.[16] Dart went on to accuse those who suggested a local origin for Zimbabwe of killing 'the investigation of Rhodesian antiquities and their history' and 'after a few further remarks, delivered in tones of awe-inspiring violence Professor Dart sat down very hard on his chair'. What is fascinating about this brief public conflict is that Dart had never seen the Zimbabwe ruins and his input was very much in the form of an armchair scientist – presenting great ideas from little data. In this he was little different from the broad white South African public in the belief that modern Bantu-speaking people were simply incapable of the creation of such wonders as Zimbabwe. After the conference, Caton-Thompson looked at the publications about the subject in the English, South African and Rhodesian press between 1929 and 1933 and found only 30 per cent of published articles supported her interpretation; the balance were either against or politely incredulous (Caton-Thompson 1983, 131).

THE ITALIAN CONNECTION

The events of the 1929 conference coincided with an opportunity for Dart to see the Great Zimbabwe ruins in what was then Southern Rhodesia at first hand. The discovery and first publication about the Taung skull had not only brought Dart international fame, but also triggered a string of visitors from the broader academic world. Many came to see the fossil and its discovery site. Among these was Lidio Cipriani of the National Museum of Anthropology and Ethnology in Florence, Italy.

Cipriani had completed his graduate studies in Florence, but this trip to Africa was his first experience as a field researcher. He was directing the scientific aspects of an expedition organised by Commander Attilio Gatti of Malan, and their destination was Zululand, where they would stay from March to November of 1927. The Zululand expedition and the subsequent longer expedition to Mozambique and Southern Rhodesia resulted in a corpus of 2 000 photographs, a series of motion picture sequences and plaster facial casts of several living Zulu individuals. The casts were part of what would become an anthropometric survey around Africa and would provide evidence to support Cipriani's racial theories (Martinaglia 2013). Cipriani had shown his Zululand face masks to Dart when they met in 1927 and at Dart's request he provided a set of these masks for Dart when he stayed in Johannesburg for the joint meeting in July 1929.

Italian physical anthropology had diverged into two distinct schools of scholarship by the 1920s. The first, based in Florence and started by Paolo Mantegazza in the 1870s, studied the skeletal morphology of aboriginal populations, with a special interest in the distribution of anthropological traits. The second, founded in Bologna by Giuseppe Sergi, but centred in the early twentieth century in Rome, was primarily focused on the development of methodology in order to produce objective and systematic classifications of present and past populations (Chiarelli and D'Amore 1997, 531). Cipriani was a member of the Mantegazza school of anthropology and had a special interest in 'exotic' peoples and their cultures in an unapologetic social Darwinian approach.[17] Ethnographic photography was an outcome of the belief that such images not only documented biology, but also reflected the expression of the intellect. Although he may not have mentioned it when he met Dart, Cipriani was a member of the Italian Fascist Party and, according to Wikipedia, was one of the most convinced supporters of the legitimacy of the colonial conquest in Africa. That Cipriani impressed Dart is not in doubt. Cipriani published in support of Dart's Taung discovery and gave him the set of Zululand face masks. The manufacture of face masks became a standard anthropological activity in Dart's department and his technician, Eric W. Williams, would use face masks to record the facial features of the Kalahari San in the 1936 expedition (Rassool and Hayes 2002, 135) and would also provide 146 Khoesan faces during the 1948 expedition to Namibia.[18] Most importantly, it was through Cipriani that Dart was invited to join the Italian scientific expedition in 1930.

The 1930 expedition that Dart joined was the third organised by Gatti, with scientific direction from Cipriani. With Zululand, Mozambique and Southern Rhodesia already completed, the third journey would encompass the northern Cape Province, South West Africa, Northern Rhodesia and the Congo (Martinaglia 2013). Gatti, Cipriani and Cipriani's scientific assistant, Nino del Grande, left

Johannesburg in September 1929 and Dart was scheduled to join them in June 1930, some eight months later. The plan was that Dart would meet the others in Lusaka, Northern Rhodesia, on 7 June and join them in the excavation of a newly discovered archaeological site at the Mumbwa caves, near Broken Hill (now Kabwe). As it transpired, Cipriani was not part of the excavation party. His interests were more ethnological than archaeological and he travelled independently through the Congo and up into the Nile Valley for his own fieldwork.

Dart travelled by train to Bulawayo in Southern Rhodesia, stopping there to take a side trip to visit the Zimbabwe ruins, and then travelled on to Lusaka where Gatti and Del Grande awaited him (Wheelhouse and Smithford 2001, 109). They travelled by car north to the Copperbelt town of Broken Hill, just north-east of the Mumbwa caves, which were their target.

At Broken Hill, the expedition was hosted by the Broken Hill Development Company, whose general manager was Royal Stevens, the father of Kathaleen Smithford, who was nineteen years old at the time. She and Frances Wheelhouse produced a biography of Dart in 2001, which is full of detailed information, but the book is difficult to use as a primary source because the views of Dart are often confused with those of the authors, but the four chapters written by Smithford provide a fascinating peek into the colonial experience that Dart was exposed to on his visit. Dart wrote almost nothing about the trip in his 1959 autobiography, but he did provide a range of reports to newspapers, including the *Illustrated London News*, the *New York Times* and *Johannesburg Star*; a published article in the magazine *The Headlight* (Dart 1930); and a long transcript of six radio broadcasts that he gave about the trip.[19] Wheelhouse and Smithford (2001, chapters 12–17) have summarised much of this material, so it is sufficient to say here that the trip involved the excavation at Mumbwa for nearly three months, as well as an extended trip through the Congo, where Dart visited colonial medical facilities, a gorilla hunt in the mountainous area of the Albert National Park and a slow journey down the Nile by rail and river boat, which included a stop in Cairo where Dart was able to see Egyptian antiquities at first hand. The only substantial scientific paper by Dart to come out of the journey was the report on the Mumbwa excavations (Dart and Del Grande 1931).

Did Dart's participation in the Italian scientific expedition influence his views about Africa in any way? It was on this trip that Dart was able to see the Rhodesian archaeological sites of monumental architecture, agriculture and mining for the first time, as well as the great sites of Egyptian antiquity. He saw gorillas in the mountains of the eastern Congo that few of his scientific contemporaries had seen. He saw the workings of the colonial administrations, especially in relation to health. He also had a break from the stress of running the Medical School in Johannesburg and the difficulties he was facing in getting the scientific world to accept his discovery at Taung. Yet his views, as seen in subsequent publications, suggest that his experience reinforced his earlier preconceptions, rather than opening new ideas for exploration. Not unexpectedly for the era, his travelling companions and hosts expressed colonial viewpoints. He was impressed with the medical missionary labours of the Catholic orders in the Belgian Congo and by the industry of the mines in Northern Rhodesia. His contact with native peoples was through the lens of white colonists, which meant that his understanding of Bantu-speaking and Khoesan peoples became locked into the mode of inferiority in which the colonial regimes placed them. Most significantly, his belief that evidence abounded of prehistoric contact between the great civilisations of Egypt and the Middle East was reinforced.

Dart's relationship with the Italian members of the expedition is also of interest. Their relationship was good except for a disagreement between Gatti and Dart at Kampala towards the end of the trip. Gatti wanted to extend the trip, but Dart was determined to begin his last leg northbound towards Egypt so that he could meet his wife in Naples (Wheelhouse and Smithford 2001, 168). There were also linguistic challenges as Del Grande did not speak English and conversations between them required the translation abilities of Gatti. Dart does not mention anything about the growing Italian desire for an African empire, but it is difficult to imagine that this topic did not come up in conversation over the months that they were together.

Dart's arrival in London in February 1931 should theoretically have been the end of the Italian expedition, but just more than two years later he received an unexpected letter from Del Grande, dated 7 June 1933, telling him that he was planning a new expedition of his own without Gatti.[20] Del Grande offered Dart the scientific leadership role if he was prepared to join this new adventure. He also asked Dart to investigate whether Wits would consider acting as a sponsor for the expedition, although he did not ask for financial support. The itinerary would be to start from Mombasa in Kenya, then on through Tanganyika (now part of Tanzania) and the Ugandan Protectorates via the Great Lakes of Victoria, Tanganyika, Kivu, Albert and Edward, then through Eastern Congo, north to Juba, Khartoum, and ending in Cairo. Del Grande listed in his letter nine scientific objectives that he hoped to achieve:

a) To do research on the life, habits and habitat of the Gorilla; Okapi; and Bongo.

b) To hunt in the Tchibinda (Kivu) forest one or more Gorillas, and possibly capture a small one for an Italian zoological garden.

c) To hunt in the Souliki Forest (Eastern Congo) one or more specimens of Okapi and Bongo.

d) Hunt one or more large Chimpanzees in Salos (Lake Kivu).

e) Hunt one or more specimens of enormous wild pigs of which very little is known, at Katana (Lake Kivu).

f) Hunt one or more specimens of large monkeys of which very little is known, at Uzania (Lake Tanganyika).

g)   Carry out ethnographic collections along the entire length of the expedition and, mainly, among the Masai, the Pygmy, the Batura, Wathusi and Wahuto.

h)   Create a series of masks of Pygmies and Batura.

i)   Gather plants, photographs, various materials for reconstruction in museum of groups of animals killed (for America).[21]

Dart contacted the university principal, Professor Humphrey Raikes, who agreed to give the expedition the backing of the university.[22] Dart wrote back to Del Grande in July accepting the offer in principle as long as he would not be required to raise funds for the trip and asking for more details.[23] Del Grande responded on 24 August with a letter full of detail. He had managed to obtain sponsorship with some promise of funding from the Fascist Institute of Culture, the Natural History Museum of Milan, the American Museum of Natural History in New York, the Cairo Museum, and the Natural History Museum and the University of Cernauti in Bucharest, Romania.[24] Most of these institutions were expecting specimens in return, so hunting and preparing specimens would be a significant part of the trip. His departure from Milan was set for 10 October 1933 and he planned to meet Dart in Mombasa on 25 November.

Although Dart had accepted the offer, he now discovered that the demands of the medical faculty would not allow him to take leave. Instead, he offered to send his new senior lecturer, Alexander (Sandy) Galloway. Galloway (1901–1965) had graduated in medicine from the University of Aberdeen in 1925 and taught anatomy at the University of Saskatchewan in Saskatoon from 1927 to 1932. According to Tobias (1968, 284), it was the discovery of Taung by Dart in 1925 that had sparked Galloway's interest in anthropology and his interest in applying for an anatomy post in South Africa. He would stay on as Dart's senior lecturer from his arrival in 1932 until 1946, with only a brief break for war service in East Africa. Galloway was an inspiring lecturer and mentor. Although nominally the science students were taught by Dart, it was actually Galloway who had the most student contact and he taught research methods and carried the anthropological torch in the Department of Anatomy while Dart carried the administrative load.

Dart introduced Galloway to Del Grande, saying that Galloway

has been with me for nearly two years and has had a considerable experience here in investigating Bushman graves and skeletons, as well as stone ruins and stone implements. He knows the technique of making plaster casts, taking anthropological measurements and, from his general experience in histology and neurology, understands fully how to preserve, pack and handle zoological material of all sorts.[25]

Dart made arrangements for Galloway to leave Durban by steamer on 30 October, to arrive in Mombasa on 14 November, where he would wait for Del Grande's arrival. Then the unexpected news came that Del Grande's father had died and that he had to postpone the expedition until early in 1934.[26] By this time Galloway was already on his way to Kenya via Dar es Salaam and he would be on his own in British East Africa until at least January, perhaps February. To his immense credit, Galloway immediately set out to make his own research objectives. He visited Dar es Salaam and Zanzibar and then made his way north to Tanga, in present-day Tanzania, where he met with the colonial secretary for native affairs.[27] He went on to Nairobi in Kenya and then back to Tanzania, to Nebarara, where there was the possibility of excavating some archaeological sites. He then travelled south to Arusha and the Ngorongoro crater with the hope of meeting Del Grande at Kigoma on Lake Tanganyika. But this meeting was not to be. In an undated letter, Del Grande informed Galloway that he could not guarantee his departure from Milan in February as his mother was very ill and that the whole expedition might have to be postponed again until 1935.[28] Galloway completed his East African exploration and gathered some useful data from archaeological and anthropological observations, which he published on his return to South Africa (Galloway 1933, 1934, 1935).

Del Grande was still trying to organise the expedition in 1934 and sent Dart a beautifully printed programme of a revised trip (which still listed Wits as a sponsor), but Dart was no longer interested in participating.[29] The new trip would no longer focus on British East Africa.[30] Instead, the trip would be to the Belgian Congo and would traverse the continent from east to west. Del Grande did not tell Dart that one of the reasons for the change in itinerary was that the colonies of Kenya, Tanganyika and Uganda had refused him permission to do research and gather specimens. This was a direct result of the political tension between Britain and Italy in relation to Benito Mussolini's expansionist interests in Abyssinia/Ethiopia. In addition, concerns had been raised about the credentials of the expedition members, stressing the need for careful scientific methods and, more importantly, that any material collected on British territory should not be allowed to fall into foreign hands (Basu and Damodaran 2015).[31]

UNDER THE SHADOW OF GRAFTON ELLIOT SMITH

Dart's arrival in London in February 1931 was an anti-climax. After eight months of travelling through Africa, he felt like a Rider Haggard character who would be welcomed as a great adventurer in his spiritual home of England (Dart and Craig 1959, 62). He felt confident that he could get his British colleagues to accept his Taung Child as a human ancestor of great importance, but almost immediately his hopes were dashed. He visited his mentor Smith as well as Keith at the Royal College of Surgeons and Arthur Smith Woodward at the British Museum. All were friendly and hospitable, but were more interested in the recent discoveries at Zhoukoudian in China than in

Dart's *Australopithecus*. Smith invited Dart to accompany him to the Zoological Society of London meeting later that month where he would be able to give an account of his fossil to the assembled scientists when Smith finished presenting his paper on the *Homo erectus* specimens from the island of Java. Dart grasped the opportunity, but with no lantern slide illustrations or casts of his specimens, his unrehearsed presentation fell flat in comparison to the masterful performance of Smith. Dart realised as he surveyed 'the unchanged expressions of my audience' that he had failed to convince his colleagues of the import-ance of his discovery (Dart and Craig 1959, 63). But there was worse news. A few weeks later, as Dart was preparing for his return voyage to South Africa, he was informed that the publications committee of the Royal Society had decided to reject the publication of his descriptive manuscript, except for the smaller section on the teeth. The descrip-tion of the Taung skull would have to rest on the exhaustive work of Keith, which recognised the importance of the discovery, but did not put it on the main line of human evolution (Keith 1931, 47).

Dart was divided in his feelings as he readied for his trip home. The failure of his British colleagues to support his conclusions about the critical nature of *Australopithecus* was extremely disappointing, but he still saw himself as part of the British mainstream and he des-perately wanted to be part of this centre of anatomical science. He knew Johannesburg would continue to provide him with unparal-leled research opportunities, but the heart of the scientific community remained in England, not South Africa. Dart then applied for the chair of anatomy at the University of Birmingham. He was supported in his application by James Wilson at Cambridge and the former prime min-ister of South Africa, Jan Christian Smuts. Smuts had been president of the S2A3 in the year of the Taung discovery and he had been one of the first people to congratulate Dart on his discovery. Smuts had a great interest in human evolution and had already published his own views on the subject (Smuts 1926) and kept in close contact with both

Dart and Drennan through the activities of the S2A3.[32] Smuts wrote in his reference letter of 6 May 1931 that Dart's 'work in anthropology has been outstanding and has secured for him wide celebrity, and with further opportunities such as a great University Centre in the Old World alone can afford, I have no doubt that he will still further distinguish himself in future'.[33] Having a former prime minister (but, in 1931, leader of the opposition) as a reference was impressive, but sadly it was not enough to sway the selection committee of the university. They instead offered the post to Robert Douglas Lockhart.

The appointment of Lockhart was a message to Dart that despite his desire to be accepted in Britain, his feelings were not reciprocated. Lockhart was almost the same age as Dart (just a year younger). He did not have the administrative experience that Dart had gained in developing the Medical School in South Africa, but Lockhart had something that Dart did not – a medical degree from a prestigious Scottish university.[34] Lockhart's alma mater was the University of Aberdeen. Its Department of Anatomy gained its first chair in 1839 and maintained a museum that contained both anatomical and anthropological specimens from the 1860s.[35] Lockhart had graduated from Aberdeen in 1918, but had stayed on as anatomist. Given the very parochial approach to British anatomy at the time, an Australian based in South Africa would not have had the same weight as an Aberdeen man – even with the support of Smuts. Dart's failure to secure the Birmingham post must have been especially demoralising, but what it did do was to convince him that his future lay in South Africa and he no longer applied for overseas posts after 1931.

The years between his return from London and the start of the Second World War were again busy ones for Dart. Beyond his duties as teacher of anatomy and dean of medicine, Dart was on the boards of the South African Institute for Medical Research, the Johannesburg General Hospital and Tara Hospital (Tobias 1984, 14). He was Wits's representative to the South African Medical Council (from 1933), a board member of the Transvaal Directorate of the Southern Cross Insurance Company (from 1934) and a lieutenant colonel and officer commanding of the First Field Ambulance, South African Medical Corps.[36] He continued to produce research despite this heavy administrative load, although his rate of publication decreased to three papers per year, but most of these papers coalesced around two focal themes that Dart had been thinking about from his arrival in South Africa: the diffusion of culture and the 'primitive nature' of the Khoesan.

Dart was unashamed of rooting his belief in the diffusion of culture in his contact with Smith (Dart 1972, 173; 1974, 160). He got to know Smith well during his stay in Smith's department in the early 1920s and Dart had planned on directing his own research into Smith's field of neuroanatomy (Dart 1972, 171). Dart's impression of Smith reflected not only his skill as a scientist, but also the strength of his personality: 'Not only was he a genius in his own field but one of the most pleasant human beings I have ever worked for or with. Tall, ruddy-complexioned and distinguished, with immaculate white hair, he was the complete antithesis of the woolly-minded, innocent genius of fiction' (Dart and Craig 1959, 29). Others who worked with Smith commented on his flexibility of mind and his ability to work with his contemporaries regardless of whether they agreed with him or not. Smith was a master at dealing with complex and conflicting information and was able to provide a 'complete and satisfying solution of the problems he undertook' (Todd 1937, 523). But not all of Smith's contemporaries shared T. Wingate Todd's interpretation of this as a positive attribute. His numerous publications 'prove Elliot Smith's intellectual stature and childlike simplicity of approach to scientific truth' (Wickham Legg 1949, 817).

Smith was a comparative neuroanatomist who became intrigued with the issue of human culture during his tenure as the chair of anatomy at the government medical school in Cairo between 1901 and 1909. It was during his time in Egypt that the first Aswan Dam was constructed, and he was given the task of examining the thousands of mummified ancient Egyptians whose remains were to be disturbed by the rising waters. Smith noted the distinct method of mummification practised by these ancient people and it struck him that there were similarities in the patterns of Egyptian mummification and those in other parts of the world. This prompted him to propose a theory that the Egyptian Cult of the Dead was the original pattern and that the practice of mummification had spread out from ancient Egypt. Smith extended his proposal to include other cultural attributes, including monumental architecture, which he consolidated into a theory in which Egypt became the seat of all human civilisation. In his words: 'The cemeteries of Egypt were truly the birthplace of the arts and crafts of civilization' (quoted in Elkin 1974, 142). Along with William H.R. Rivers and William J. Perry, Smith became one of the leading proponents of the British school of diffusionism.

The great debate about independent invention or cultural diffusion was part of late nineteenth-century European anthropology (Penniman 1952, 176) and was linked to the rise of social Darwinism as a scientific model for the origin of cultures. Strict social Darwinists believed that human cultures underwent an evolutionary progression and that primitive races were at lower stages in the process of moving towards the development of higher culture, of which Europeans were the epitome. The diffusionists, on the other hand, believed higher culture was the property of more advanced people and that cultural elements diffused outwards from these progressive societies and the superior traits were borrowed and adapted to the local context (Kuper 1983, 3). Primitive races were simply that – primitive. The debate gained complexity in the early twentieth century with the introduction of the functionalist and structural-functionalist schools of social anthropology by Bronisław Malinowski and Alfred Radcliffe-Brown, respectively. Social Darwinism and diffusionism both concentrated on the material culture and political structure of society and its progression over time. Social anthropology viewed social phenomena as dynamic and able to adapt to local conditions. In practice, this view of human culture was ahistorical and it concentrated on the process of social development, not its history (Kuper 1983, 5).

The social anthropological perspective took root in the developing field of anthropology in early twentieth-century South Africa. Radcliffe-Brown launched the School of African Life and Languages at UCT in 1921 and returned to South Africa late in his career to start a similar programme at Rhodes University. Winifred Hoernlé began teaching ethnology at Wits in 1923, but she collaborated closely with Radcliffe-Brown in Cape Town (Carstens, Klinghardt and West 1987, 8). Nearly all of Radcliffe-Brown's and Hoernlé's students were to take Malinowski's anthropological seminar series at the London School of Economics after 1920, which included attacks on diffusionism (Kuper 1983, 26; Dubow 1996, 18).

Dart's approach to questions of the origins of both people and their cultures in southern Africa was an inflammatory mix of typology and diffusionism. Dart's reaction to Caton-Thompson's paper at the 1929 meeting should not be surprising, given the strength of his support of Smith's theories. In one of his 1929 S2A3 papers, Dart proposed that detailed anthropological studies of the South African Bantu-speaking peoples would support 'my belief concerning Egyptian, Semitic, Arabic, and Mongoloid infiltrations into the population' (Dart 1929c, 315).

Dart's reputation for supporting Smith's hyper-diffusionist model was well known, right from his earliest days in Johannesburg. Writing to her former Cambridge tutor Alfred Cort Haddon on 28 April 1926, Hoernlé expressed her concern that Smith's diffusionism might gain too much influence at Cambridge and she noted that Dart was doing just that in Johannesburg. She wrote to Haddon:

Professor Dart is a great trial in many ways, for he is a keen Elliot Smith man and is more keen than thorough. He starts all sorts of hair brained [*sic*] theories without any basis or practically none, and of course captures the popular imagination with the result that one is kept busy trying to check the students from doing the same thing! (quoted in Carstens, Klinghardt and West 1987, 194)

Dart's research in the 1930s followed on from the earlier papers he had written about Boskop. As an alternative to, and earlier version of, the 'Bushman type', efforts to identify Boskop features in archaeological skeletons became routine. Dart's chief acolyte in this was Galloway (Galloway 1937a). Dart was asked to analyse the skeletons when the gold graves were discovered on top of Mapungubwe Hill in December 1932. The pressure of work and a fear that his dispute with Caton-Thompson would result in his work being regarded as prejudiced prevented him from accepting the offer (Dubow 1996, 17), and he recommended Galloway instead (Fouché 1937, 126). Initially, the job was substantial, but it soon became huge. The hilltop at Mapungubwe produced 24 skeletons, but they were in poor condition and bones from only 11 individuals could be analysed (Galloway 1937c, 127). When the excavation moved to the neighbouring site of Bambandyanalo in 1934, more than 70 graves were identified and the skeletons were in general in much better condition, providing a large sample of 40 individuals for analysis. Galloway was diligent, but he needed the help of Lawrence Wells to complete the job. Wells, still studying for his medical degree at the time, was a demonstrator in anatomy in the department, but was extremely keen to be involved in research. Galloway and Wells completed their task in 1938, with Galloway concluding that the people buried at Mapungubwe and Bambandyanalo were 'a homogeneous Boskop-Bush population' of 'pre-Negro' origin and that there was no link between them and the Bantu-speaking peoples of southern Africa (Galloway 1937c, 162; 1959, 125). Although neither Dart nor Galloway made the formal connection in writing, this conclusion rested firmly in the white settler ideology that Bantu-speaking peoples

Figure 5.2: Gold artefacts were discovered with the Iron Age burials from Mapungubwe Hill in 1933. Raymond Dart was offered the task of analysing the skeletons in 1935, but recommended his senior lecturer Alexander Galloway for the work instead. Galloway drew in the science students under Lawrence Wells to help with the project when more skeletons were excavated from the nearby site of Bambandyanalo. This photograph shows (from left): Lawrence Wells, Alexander Galloway and Trevor Trevor-Jones studying the Bambandyanalo skeletons in 1936. Trevor-Jones and the photographer, G.S. Chenik, were third-year science students at the time. (Source: Dart and Craig 1959, photograph by G.S. Chenik)

had not attained a level of civilisation that would have enabled them to build such marvellous sites (Hall 1990, 61).

In the conclusion of his 1937 chapter in Leo Fouché's book on Mapungubwe, Galloway expressed regret that the preparations for the Kalahari expedition prevented Dart from giving him help and criticism on his draft (Galloway 1937c). Although Dart was only indirectly involved in the discussion of the remains from Mapungubwe, he was directly involved in the planning and execution of the Kalahari expedition to Twee Rivieren near the border with South West Africa. The story of Donald Bain and the group of San people that he brought to Johannesburg and Cape Town has been well told by Cirraj Rassool and Patricia Hayes (2002, 129) and they, along with Christa Kuljian (2016, 61), have noted Dart's role in the project. The opportunity to study a group of 'pure Bushmen' first in the Kalahari and then for an extended period at the Frankenwald farm in Johannesburg was unprecedented. The scientific team under Dart published two volumes of reports in the journal *Bantu Studies*, covering language, music, games and dances, genealogy, diet, technology and health and disease, but Dart himself contributed an extensive chapter on the physical characteristics of the group (Dart 1937a). Dart's effort was a mastery of typological assessment using non-metric characteristics. These characters were based on the Italian system of cranial shape and facial feature descriptions, which translated continuous morphological variation into discrete categories. This enabled him to describe racial origins in terms of specific features and therefore to assess multiple genetic lines in each individual. Dart identified 'Bush' and Boskop types (or intermixtures between them) in his group of San. Features that did not fit either of these predefined groups were assigned to 'alien admixture', according to Dart's preconceived diffusionist ideas: 'The alien admixture was shown to be principally Brown (or Mediterranean or Hamitic) but in addition Mongolian and Armenoid and only to a minimal extent Negro' (Dart 1937a, 227).

WELLS, GALLOWAY AND THE LASTING INFLUENCE OF DART'S TYPOLOGY

When the Second World War broke out, Dart's role as lieutenant colonel in the South African Medical Corps became a heavy load. He now carried the training of medical staff for war duty as well as his duties as dean and professor in the Medical School, but his personal life also took a difficult turn. His second child, Galen, had been born with several developmental problems and Dart took it upon himself to help the child with his physiotherapy. Together, these problems were too much and Dart suffered a physical and nervous breakdown in 1943. He was ordered to relinquish all duties for a year and to rest (Wheelhouse and Smithford 2001, 221).

When he returned to the Department of Anatomy in 1945 Dart had made several changes to his lifestyle. He was no longer dean of the Faculty of Medicine and he began to delegate his departmental duties and teaching. Hertha de Villiers remembers when she joined the department in the early 1950s that Dart did not like staff meetings, so every day,

> he would start off in the basement and go to the top and see virtually everybody. He always had a briefcase with him, and in this briefcase invariably he had something for you, either a letter he had got or an article he'd written because everybody on the staff including technicians had to have a project so that they were committed so it wasn't just an 8 to 5 job.[37]

This enabled Dart to concentrate on research, something that was immediately reflected in his publication output. Between 1946 and his retirement at the end of 1958, Dart produced an average of more than five papers per year, but this accelerated in the first ten years of his retirement to nearly ten papers annually.

What gave Dart the ability to focus on research was the presence of two highly competent and energetic staff members who could help to carry the load. First of these was Galloway, who had been Dart's senior lecturer from 1932, and second was a home-grown anatomist, Wells, who had been a mainstay as a researcher from his student days in the department. It was Wells who would become one of Dart's most fervent typologists, developing his views not only from Dart, but also from Galloway. When Wells left for Edinburgh in 1951, his post would be taken by Phillip Tobias, who would find his own way to mark his place in South African physical anthropology.

Wells was born in England in 1908 and immigrated to South Africa with his parents in 1920. After high school in Johannesburg, Wells took the science course in anatomy under Dart, graduating with a Bachelor of Science degree in 1928 and a Master's degree two years later. He continued his academic career at Wits, acting as a demonstrator and junior lecturer in anatomy while he studied for his MBChB, which he obtained in 1938. Choosing not to practise clinical medicine, Wells stayed on in the department, being promoted to lecturer in 1941, senior lecturer in 1945 and finally a special position as reader in physical anthropology in 1948 (Tobias 1988, 12; Štrkalj 2006b).

Wells was a consummate researcher, writing his first paper on anatomical variation while still an undergraduate student. Dart soon roped him into working on the archaeological skeletons coming into the department. Wells described the remains from two new sites disinterred by FitzSimons: Zuurberg, north of Port Elizabeth, and Whitcher's Cave, near George on the Cape south coast. Along with Jock Gear, Wells also did the analysis of the skeletons from a more systematic excavation by Clarence van Riet Lowe on two isolated cairns from near the Orange River. His contact with Van Riet Lowe exposed Wells to serious archaeological research and this would become a long-term interest of his, extending his work beyond skeletons into pottery, stone tools and, most extensively, into faunal analysis. In parallel with his archaeological work, Wells concentrated on human structural anatomy, completing his Master's degree on 'the foot of the South African native' in 1930. It was published in full in the *American Journal of Physical Anthropology* in 1931. He would continue this detailed anatomical research for his 1946 doctorate on the function and phylogeny of the human leg and foot (Trevor-Jones 1980, 261). These varied early research projects all show the hallmarks that would be a sign of Wells's research throughout his career. His work was done in meticulous detail and even if his typological conclusions were not accepted in the long term, there were very few speculations and his inferences were always based on the data observed. Wells was not lured by Dart into Smith's theories of diffusion. In Tobias's words, Wells 'was a cautious scientist, perhaps in some respects the antithesis of his mentor, Professor Raymond Dart' (Tobias 1988, 13).

Wells's primary role in much of the Department of Anatomy's typological research was to gather data. He cleaned, reconstructed and measured skeletons for Galloway's Bambandyanalo project, as well as several other skeletons that were brought into the department through chance discoveries or from archaeological excavations. He became involved in a series of expeditions to gather anthropometrics of living Khoesan and Bantu-speaking populations. In April and May of 1934 Wells visited Christiana and Bloemhof in what was then the Western Transvaal to measure 'Korana', followed by a trip to Windhoek in 1937, when he measured twelve 'Hottentot' men in the Windhoek Location and eight more in the Windhoek jail.[38] None of these data were ever published and it is unclear why the expeditions took place at all. One possibility is that they were follow-up trips from an earlier expedition led by Professor Louis Maingard of the Department of French and Romance Languages, who had a special interest in the languages of Khoe and San peoples (Murray 1982, 143). Maingard visited the Korana near Bloemhof in February 1932 with his son, J.F. Maingard. While Maingard senior gathered information on customs and language, Maingard junior took the opportunity to gather some anthropometric data on eleven individuals. Father and son published their papers in 1932 and Maingard junior thanked both Dart and Wells for the instruments and advice (L.F. Maingard 1932; J.F. Maingard 1932). The published data were very limited and it is likely that Wells planned his own trip to gather more reliable data on the same people at Bloemhof and neighbouring Christiana in 1934, collecting measurements and genealogical data on 32 people.

The 1937 trip to Windhoek was a more opportunistic outing as it coincided with the S2A3 meetings for that year. Galloway had already been to Windhoek two years previously and had measured Ovambo people (Galloway 1937b). It was around this time that Wells was drawn into a project measuring African women at the Bridgman Memorial Maternity Hospital in Johannesburg. He and fellow student Margaret Orford were asked by Dart to look at the anthropometry of women patients with the overall objective of gathering data on the size and structure of the pelvis in black women. Catherine Burns (1996, 10) has been highly critical of this research for just reasons. There was no evidence that patients were informed about the project or why they were being measured and the measured assessments went far beyond pelvic dimensions. Orford and Wells included data on skin colour, facial features, buttock projection (steatopygia) and the tablier (elongation of the labia minora). The published study (Orford and Wells 1937) was more concerned with obtaining the definition of the racial type than being a guide to obstetric outcomes. Burns suggests that this was part of a programme of field studies on living peoples inaugurated by Dart, but the project was in fact not part of any organised programme other than a reflection of Dart's vision of typology in a broad definition of past and present peoples in Africa. In fact, one of Dart's legacies was the development of multiple focuses in the science course: from skeletal biology and palaeoanthropology through primate studies and comparative anatomy to clinical anatomy and pathology.[39] Burns also suggests that Orford and Wells continued their study in northern Natal in the late 1930s, but there is no evidence of this in any of Wells's papers nor is it noted in the Tobias paper to which Burns refers (Tobias 1970).

Wells would undertake one more expedition to study living people in 1947–1948. He and Eric W. Williams, the senior technician in the Department of Anatomy at Wits, would join the University of California African Expedition to measure Ovambo and !Kung in eastern Ovamboland in South West Africa and adjacent Angola (Camp 1948, 550). Wells collected physical anthropometric data and Williams made face masks. Wells saw this not as an opportunity to generate a 'Bushman type', but as an opportunity to study the variation in a population of San who could be compared to other San groups (Wells 1952, 54). This would be among the last papers that Wells would publish that focused on South African 'native types', living or skeletal, as by the early 1950s his interests had taken an entirely different path.

Wells did not enjoy working with living peoples and his favourite research involved laboratory or library studies. He was an intensely shy man and he took a long time to obtain his medical degree because he struggled to cope with gynaecology and obstetrics.[40] Both Alun Hughes in Johannesburg and Ralph Kirsch in Cape Town remember him walking down passages at their respective medical schools with his hand touching the wall.[41] Although he took on a substantial administrative load in both Johannesburg, as a senior lecturer, and in Cape Town, as a professor, he avoided staff meetings and filling out official forms as much as possible. Hughes remembered that when Wells oversaw the second-year medical students in the late 1940s, he would give Hughes a single sheet of foolscap paper on which a six-day weekly schedule was drawn: 'He would hand it to me and he'd say, "Get the secretary to type two copies, put one in each dissection hall, and give me this one back" and that was all the paperwork for twelve months.'[42] Although deeply interested in the museum collections, Wells was not consistently good at keeping records or building the collections. No cataloguing of the archaeological skeletons was done and no new skeletons were accessioned to the research collection of cadaver skeletons during the period that he was head of department in Cape Town.

Wells left Johannesburg for the post of reader in physical anthropology at the University of Edinburgh in 1951, where he stayed until taking up the post of professor of anatomy at UCT in 1956. His sojourn in Scotland saw a change in his skeletal biology research focus, but it is the period after his return to South Africa that marked a significant change in the nature of his research.

Wells's appointment in Edinburgh was as a physical anthropologist, not an anatomist. With few anatomical duties to perform, Wells chose to look at the British archaeological skeletons available to him. The anatomical museum at the Edinburgh Medical School had a large collection of pre-Roman skeletons excavated from archaeological sites in Scotland. Wells began a systematic project of measuring each cranium, with the idea of producing a definitive work on the physical anthropology of Bronze Age Scotland. At the same time, he took on the role of osteologist for British archaeologists and ultimately completed nine reports on sites from the Shetland Islands in the north to Dorset in the south of Great Britain. He did not manage to complete his grand project by the time he returned to South Africa in 1956, but stored in the files of the Department of Human Biology at UCT are seven drafts of various papers on the subject, with 34 pages of tables of cranial measurements.[43]

Comparing Wells's published output before and after 1950 is very instructive. Before 1950 he published sixteen reports on archaeological skeletons. In the fashion of Dart, these concentrated on age, sex and racial typology, with little or no information about the life experience of the individuals concerned. He was also intrigued by archaeology, publishing fifteen archaeological papers – all published before he left for Edinburgh. During the 1930s and 1940s Wells began to develop an interest in faunal analysis and wrote short papers on a range of fossil and sub-fossil animal bones – from baboons and hyraxes to giraffes and horses. These included two papers on fossil bones from the australopithecine excavations at Makapansgat in 1947 and

1948. But it was as if a floodgate opened on the topic after 1950, with Wells publishing a total of 23 more papers, mostly about antelopes. After 1950 Wells also became more interested in the broader picture of human evolution, rather than the local South African context. By the end of his career, he had written nineteen papers on the subject. He also produced another nine skeleton reports, but these were all on more ancient skeletons that concerned palaeoanthropology, rather than more recent skeletons. Older academics often develop an interest in the history of their field and this was certainly true for Wells. He wrote several historical notes about early documents that he had come across, including the documents about the donation of 'Bushman' skeletons that he found in United Kingdom collections, but his real late-career interest was in medical history. Between 1959 and 1975 he published ten papers on medieval and Renaissance medical illustration (Morris and Keen 1982, 42).

There were probably several reasons for Wells's shift in interests after he left Johannesburg. He had been Dart's student and colleague for his entire academic career and to a large extent he had been working on subjects suggested and supervised by Dart. He did not follow Dart's diffusionist agenda and his archaeology and faunal analyses were meticulous and independent, but it is in the description of human skeletons from the various site reports that we can see Dart's typological influence. These were the studies that stopped once Wells had left the Johannesburg fold. Was Wells overwhelmed by Dart? Ralph Kirsch and Ronald Singer, who knew him during his time in Cape Town, thought that might be the case and Wells himself recognised how much influence Dart had had on his career.[44] Writing to Dart immediately after graduating with his doctorate in 1946, Wells said:

> I realise that I owe to you both the inspiration and the opportunity of working for this degree, that my choice of subject was the result of your guidance, and that the completed work owed a great

deal to your detailed criticism of the draft as well as to our many discussions of the problems touched upon in the thesis.[45]

But in the same letter he gave a little hint that he did not always toe the line: 'May I add that even when we do not entirely agree with each other, I can never imagine any Departmental Head whom I would have been happier to work under as student and as assistant.' Just before he left Johannesburg, Wells began work on a book that, if published, would have been a landmark for craniology. It was titled 'The Skull of the South African Negro'. Wells had gathered 147 skulls for analysis in two groups of roughly equal numbers: Zulu and Basotho from the Department of Anatomy's cadaver collection and Ovambos, loaned to him from the South African Museum in Cape Town. His plan was to compare eastern and western people of South African Bantu-speaking origin, using the same typological assessment techniques that the department in Johannesburg had been using for the past twenty years. He never finished the work and the manuscript stayed with his papers in Cape Town after his retirement.[46]

Galloway remained in the shadow of Dart during his time at Wits. He had grand plans to develop physical anthropology in Uganda when he took the chair of anatomy at Makerere, and he did collect some archaeological skeletons, which were kept in the Department of Anatomy (Obituary 1965, 748). Sadly, the demands of being dean prevented Galloway from continuing his anthropological research and his time was overwhelmingly spent on building the new medical faculty, its library and the Uganda Foundation for the Blind. Galloway was well remembered in Johannesburg (Tobias kept a photograph in his office of Galloway in full Scottish regalia), but his long-term impact was limited. So too did Wells leave little of his own presence when he retired from UCT in 1972. Although he taught physical anthropology to the students in the Department of Archaeology, he had few of his own students in the science course in anatomy. He took over the administration of the Hopefield fossil site from Drennan, but did not become directly involved in the excavations. Singer was impressed with his knowledge of bovids, saying, 'He knew his antelopes, I'll say that for him because I knew that when he came to Cape Town, he began looking at the Hopefield stuff. He really knew his stuff. In fact, he taught me a lot about antelopes which I didn't know about.'[47] However, apart from the 173 scientific publications he produced during his career, Wells left little in terms of a legacy to the department.

THE RACIST IMPLICATIONS OF DART'S TYPOLOGY

Galloway and Wells were followers of typology, but their research kept to descriptions of the skeletons and there is little speculation in their writings, unlike in the case of Dart. Of all of Dart's papers published in the years between the First and Second World Wars, his 1939 paper on 'Population fluctuation over 7 000 years in Egypt' is not only the most complex, but also shows how typological assessment and diffusionist models were mixed and closely linked in his thinking and how his speculations went far beyond the evidence.[48] Relying heavily on Smith's work on Egypt (Smith 1923) and Sergio Sergi's classification of cranial forms, Dart created a sweeping statement of successive invasions all marked by changes in cranial shape.[49] He set out a classification of nine intermingling races: two 'Caucasoid' (the Nordic and the 'Armenoid'), who 'have given the world most of its adventurous acting and thinking, invention and philosophy, both destructive and constructive'; the brown-skinned Hamitic or Mediterranean; the Mongol or 'yellow' race, who biologically 'are dominant types, and are not burdened with a sense of inferiority'; the 'juvenile, and less prolific, active and industrious, but far more light-hearted and happy Negro type of mankind'; two races with more adult cranial features, the Boskop and the Australoid; and two pygmy races, the

'Bush' of Africa and the 'Negrito' of the Andaman Islands. These last two Dart regarded as child-like in cranial form, saying that they were characterised by 'their trustful, infant outlook' and their 'merry, dancing care-free life. They are the children of men, the prototypes of fairies, gnomes and pixies' (Dart 1939, 96–97). In total, there are only two themes in this entire paper: (1) rigid races with a hierarchy of behaviour and culture; and (2) a flow of culture and civilisation from the 'more developed' races to those that are 'primitive'.

Despite the changing world view on the meaning of race, Dart's biographers (Wheelhouse 1983; Tobias 1984; Wheelhouse and Smithford 2001) did not raise the obvious issue of the racist paradigm reflected in his writings. Tobias refers to his 'tendency to overstate the case' and argues that Dart's tendency to make unsubstantiated claims reflected the freeing of his mind 'from the shackles of authority' (Tobias 1984, 51). Wheelhouse, in two volumes of hero-worship, states that the 'hallmarks of Dart's work were his objectivity, his prescience, his ability to formulate theories, and in time, prove most of them' (Wheelhouse and Smithford 2001, 183). One could argue that Tobias was protecting Dart's reputation because it was linked to his own, but it is difficult to understand Wheelhouse's uncritical adulation, as her work formed the basis of a doctorate from the University of Sydney in Australia.

Did Dart understand that his writings had racist implications? Saul Dubow (1995, 45) notes that Dart's focus on typology was used by others to reinforce racial inequality, although Dart himself did not appear to have held conventionally racist views. When his views on the Zimbabwe ruins and the skeletal interpretations of Mapungubwe and Bambandyanalo disagreed with the archaeological data, Dart (along with Galloway) saw physical anthropology as having primacy over archaeology and he considered the failure to ascribe the ruins to the activities of local Bantu-speaking populations as being perfectly justified because of this. When he gave a lecture tour of Rhodesia in June

1935, a local archaeologist wrote to the *Bulawayo Chronicle* expressing his dismay that someone with as little archaeological experience as Dart could be dismissive of the work of Caton-Thompson and others (Dubow 1996, 23). Dubow comments: 'For a scientist with an international reputation, his indiscriminate willingness to entertain diffusionist theories based on the slimmest of evidence is remarkable' (1996, 20).

Much of Dart's ability to publicise his ideas came directly from the international reputation that the Taung discovery had given him. In the same manner as Smith, Dart's talent at public speaking and his facility in popular writing made him visible to the South African public. His public audience in the 1930s was anxious about racial hybridisation and miscegenation and his writing and public lectures about race origins and purity of type reinforced racist public opinion (Dubow 1996, 28). When he spoke or wrote about 'Bushmen' as living remnants, highlighting the difference between 'primitive' and 'civilised' and the exclusion of Bantu-speaking peoples from the making of Mapungubwe and Great Zimbabwe, this helped to confirm the public belief at the time that black people were incapable of civilisation and were unsuitable for citizenship in a common society.

It is difficult to find a record of Dart's personal political views. Other than his statement about the racial situation in Europe in his paper on 'the present position of anthropology in South Africa' (1925b), Dart has left few comments that can be interpreted in any meaningful way. He was broadly supportive of segregation in the 1930s, supporting the idea of a separate medical school for 'non-white' students in his role as a member of the Loram Committee, and in supporting Bain's call for a separate 'Bushman' reserve. In the first case he was adamant that black African medical practitioners should have the same quality of training as their white counterparts (Murray 1982, 303) and he was closely involved in the Bridgman Memorial Hospital's efforts to improve standards of gynaecological and maternal care for African women (Dubow 1996, 10), while in the latter case he felt that segregation would help in the conservation of cultural heritage, as this was being destroyed by 'Europeanizing the native' (Kuljian 2016, 61). Such beliefs were not out of line with prevailing progressive attitudes in the 1930s. Well into his retirement Dart was asked to take on the role of honorary president of the Association of Medical Students of South Africa. In his letter acknowledging Dart's acceptance, the student president, Peter Arnold, made it clear that the organisation was non-racial and that they were disappointed that the Afrikaans-medium medical schools had withdrawn their membership, so clearly Dart supported this position.[50] Dart also was happy to congratulate Helen Suzman when she was re-elected to her position as the sole member of the opposition Progressive Party to sit in Parliament.[51] Hughes, who spent many hours in private conversation with Dart during his tenure as chief technician in the Department of Anatomy, noted that Dart's political views were liberal, but he never talked about them in public: 'He talked to me every day for twelve years, but never about politics. Dart was liberal at heart, but he never discussed politics, ever.'[52] Dart probably regarded himself as a political moderate (Dubow 1996, 10).

Kuljian (2016, 72) has identified the contradiction between Dart's call for respect and dignity for the 'Bushmen' as human beings and his treatment of the same people as specimens in his scientific research, especially in relation to the acquisition of the body of /Keri-/Keri, one of the individuals who had been part of Bain's exposition.[53] In fact, Dart made a marked effort to separate his science from broader society. Even when it was evident that the two intersected – for example, on the issue of race and social policy – Dart refused to make the connection. In this respect, he subscribed to what was considered the ideal ethos for a scientist in the mid-twentieth century. Science was supposed to be dispassionate and apolitical (as opposed to the engaged and politicised humanities) and the belief was that science transcended and was above morality (Marks 2009, 74). Dart simply refused to see the social and political implications of Smith's diffusionist model and this allowed him entirely free rein in the exploration of untested wild claims and rigid typological definitions (Derricourt 2009, 278). His use of typology gave each racial construction a well-defined identity and it became a simple matter to link his biological types to cultural behaviour or archaeological remnants. Ironically, Dart's clinging to Smith's diffusion did the opposite of breaking with authority. What began as a radical approach to issues in African prehistory became an inflexible adherence to the diffusionist model. Dart's reputation from Taung gave him a fast track for publishing wild and poorly thought out ideas. Robin Derricourt (2009, 281) goes further and puts many of Dart's wilder claims into the realm of pseudoscience. Dart's typology and diffusionist arguments were received well in South Africa because 'South Africa was receptive to ideas that would not challenge the racial categories that reinforced perceptions of power and difference' (Derricourt 2009, 276). Dart felt that he could comfortably distance himself from the most extreme racial paradigms because he did not deal in politics.

Did Dart's avoidance of political issues mask a more serious prejudice? When does neutrality become blindness? Dart touched shoulders with at least two outright race scientists during his career – Lidio Cipriani and Reginald Ruggles Gates. It is theoretically possible to argue that Dart was unaware of Cipriani's politics, but this is extremely unlikely. Dart was a corresponding member of both the Italian Institute of Palaeontology and the Italian Society of Anthropology, Ethnology and Psychology. Nearly all of Cipriani's work was published in Italian, but it was very public. Cipriani saw his facial casts as scientific documents and displayed the casts at conferences in London and New York (Piccioni 2020, 27). It is almost inconceivable that Dart would have spent months with Del Grande without them speaking about Italian nationalism and racial evolution in Africa and the role of Italian scientists. Dart and Cipriani exchanged reprints of their publications. In 1938 Cipriani became a signatory of one of the most racist documents in modern European history. He, along with eight other prominent Italians, published a *Manifesto on Race*, which not only positioned modern Italians as members of the high Aryan race, but relegated Jews and Africans to the level of subhumans (Pankhurst 2005, 149). This manifesto was published just over a year after the first official Italian government racist decree prohibiting conjugal relations between Italian citizens and the colonial subjects of the new Italian East African empire.

However unlikely it might be that Dart was unaware of Cipriani's views, the same cannot be said about Gates's views on race. Gates, a botanist by training, was among those geneticists who applied a simplified version of Mendelian genetics to humans (Gates 1929). By the late 1940s, Gates had proposed that human races were diverging into different species, but his proposals had been rejected by geneticists and non-geneticists alike (Weiner 1952, 407). Gates first wrote to Dart in April 1946.[54] He was working on his upcoming book about human ancestry (Gates 1948) and wanted to include some of Dart's pictures. When the book appeared, the reviews were scathing and Gates referred to them in his next letter to Dart. He wrote:

> Of course the 'chosen race' have already attacked my book over here. Here, as in England, they try to suppress all scientific discussion of race, so as to leave them free to pretend that racial differences (except the chosen race) scarcely exist, which I think is the height of hypocrisy. They play in with the Negroes and, having more brains, lead the Negroes in racial propaganda.[55]

He went on to label critics who were not Jewish as 'camouflaged Jews', but saved his greatest vitriol for Ashley Montagu, who in his mind represented the mindset of American Jewish scientists. He summed up by saying: 'The Jews in this country thus have a pretty free hand to make any attack they like and get away with it.'[56] Dart's response completely ignored the racial slurs. He congratulated Gates on his new book, a tour de force, and said how important it was that Gates had presented Fabio Frasetto's work (on cranial shape) in English and how pleased he was that Gates had given such prominence to 'the South and East African workers'.[57] Dart was to visit Gates in Massachusetts later in 1949 and hosted him in Johannesburg in 1955.

The Dart–Gates correspondence in the late 1940s occurred when Gates's reputation as a scientist was beginning to fail. As a scientist who had spent his whole career within the scientific mainstream, Gates found his level of isolation and the hostility shown to him by other scholars distressing and infuriating (Schaffer 2007, 262). He was financially supported by American segregationists, who agreed with his argument that racial mixing was disastrous to the mental and physical health of their offspring. When, in 1947, he was fired from Howard University in Washington for using his black students as anthropological data, his complete lack of political awareness made him fail to understand that such studies were now perceived as contentious. He took the criticism of his new book *Human Ancestry* as a direct political attack on his studies, and when prestigious scientific journals, including *Science*, the *American Journal of Physical Anthropology* and the *Journal of the American Society of Human Genetics*, repeatedly refused to publish his work, Gates engaged in acrimonious correspondence with their editors, accusing them of being part of the Jewish and black racial cabal against him (Schaffer 2007, 265). Gates saw himself as a balanced and neutral researcher who was under attack from non-scientific and politically motivated individuals. Despite being an anti-Semite, a supporter of American segregation and South African apartheid, Gates insisted he was an objective scientist and not a political protagonist.

Why did Dart ignore Gates's blatant racism? Perhaps he shared Gates's views on Jews and blacks, but on balance this seems unlikely. Dart wrote a glowing obituary for Gates in the pages of *Mankind Quarterly* (Dart 1966), a journal that Gates had helped to launch and which openly courted the apartheid regime in South Africa and vigorously defended American segregation. The views of *Mankind Quarterly* jarred glaringly with Dart's presumed liberal views, but were also completely opposed to the very public position of Dart's successor, Phillip Tobias. Dart seemed to be completely unaware of the political consequences of his scientific viewpoint and how it supported the unsavoury aspects of Gates's associations. What is evident is that the two researchers shared an unfailing belief that politics was separate from science and that their work was completely objective. It did not help that, from an academic perspective, Gates supported Dart on two issues: he wholeheartedly acknowledged the pivotal importance of the Taung Child in human evolution and his uncritical and simplistic typology matched that of Dart's own beliefs.

Of all those who studied racial variation in the period before the Second World War, Dart was by far the most tenacious in his belief in the link between race and human behaviour. He strongly influenced those in his academic orbit, especially Galloway and Wells. Only Wells seems to have tried to break away from the Dartian typological strait-jacket, but the evidence of his unpublished South African craniology manuscript suggests that he was unsuccessful. In a sense, Wells was a failed transition from the old typological anthropology of the first half of the twentieth century to the new ideas of the 1950s and 1960s. In the end, it would be the new generation of anatomist-anthropologists, Singer and Tobias, who would succeed in making the change from the old to the new.

NOTES

1   Alun R. Hughes, 'A Method of Cataloguing an Extensive Anatomy Teaching and Research Collection', unpublished typescript, dated 1969, on file in the Department of Anatomical Sciences, University of the Witwatersrand, Johannesburg.

2   Hughes, 'Method of Cataloguing', 1.
3   Department of Human Biology, University of Cape Town, letter from Gordon Laing to Matthew Drennan, 23 March 1938.
4   University of the Witwatersrand Archives, Johannesburg, letter from Winifred Hoernlé to Raymond Dart, 8 February 1927.
5   University of the Witwatersrand Archives, Johannesburg, letter from Tom Barnard to Raymond Dart, 11 September 1927.
6   University of the Witwatersrand Archives, Johannesburg, letter from John Phillips to Raymond Dart, 22 October 1927.
7   University of the Witwatersrand Archives, Johannesburg, letter from Dart to Barnard, 3 November 1927.
8   University of the Witwatersrand Archives, Johannesburg, letter from R.F.A. Hoernlé to Raymond Dart, 10 November 1927.
9   Three Gear brothers would take the science course exploring anthropological or primatological topics under Dart. All three would continue in medicine and have distinguished medical careers: Harry Sutherland Gear (MBChB, 1929) in public health, James Henderson Sutherland Gear (MBChB, 1930) in polio research and John Hallward (Jock) Gear (MBChB, 1933) in tropical medicine.
10 Hughes, 'Method of Cataloguing', 1.
11 Hughes, 'Method of Cataloguing', 2.
12 Tobias is adamant in linking this collection back through Terry to Turner in Edinburgh (1984, 7; 1987, 32). The collection of cadaver skeletons, along with their demographic and medical history details, was Terry's model that Dart adapted for South Africa and there is little in common between the Edinburgh collection and that of Terry or Dart.
13 Dart provides detailed instructions for embalming bodies in his letter to Dr Louis Fourie (Wheelhouse 1983, 36).
14 Hughes, 'Method of Cataloguing', 2.
15 A complete list of Dart's publications from 1920 to 1974 can be found in Wheelhouse and Smithford (2001, 321).
16 'The Zimbabwe Mystery, Prof. Dart Makes Heated Speech', Cape Times, 3 August 1929.
17 Social Darwinism is the cultural analogy to biological Darwinism, arguing that cultures and nations compete against one another in the same way that species do. The theory reached is heyday at the end of the nineteenth century. Although it was discredited in the twentieth century, it remained influential in political theories, especially in the German- and Russian-speaking world.
18 Hughes, 'Method of Cataloguing', 3; Department of Human Biology, University of Cape Town, Wells data, 'Face Masks: South West Africa 1948'.
19 University of the Witwatersrand Archives, Johannesburg, Raymond Dart, 'South to North through Africa', a series of radio broadcast lectures, 1931.
20 University of the Witwatersrand Archives, Johannesburg, letter from Nino del Grande to Raymond Dart, 7 June 1933.
21 University of the Witwatersrand Archives, Johannesburg, letter from Del Grande to Dart, 7 June 1933.

22 University of the Witwatersrand Archives, Johannesburg, letter from Humphrey Raikes to Nino del Grande, 27 July 1933. Although Dart was the intermediary, Raikes responded directly to Del Grande.

23 University of the Witwatersrand Archives, Johannesburg, letter from Dart to Del Grande, 2 August 1933.

24 University of the Witwatersrand Archives, Johannesburg, letter from Del Grande to Dart, 24 August 1933.

25 University of the Witwatersrand Archives, Johannesburg, letter from Del Grande to Dart, 26 September 1933.

26 University of the Witwatersrand Archives, Johannesburg, letter from Del Grande to Dart, 8 November 1933.

27 University of the Witwatersrand Archives, Johannesburg, letter from Alexander Galloway to Raymond Dart, 15 November 1933.

28 University of the Witwatersrand Archives, Johannesburg, letter from Nino del Grande to Alexander Galloway, November 1933.

29 University of the Witwatersrand Archives, Johannesburg, letter from Dart to Del Grande, 24 September 1934.

30 University of the Witwatersrand Archives, Johannesburg, 'Spedizione Nell'Africa Centrale: Traversata del Continente Nero da Est a Oust sulla Linea dell'Equatore – Oranizzata e Comandata da Nino del Grande, Febraio–Agosto, 1935'.

31 The central concern was that no one in the party had formal archaeological training. Del Grande was a structural engineer. He had done most of the excavation at Mumbwa, meeting with Dart and Gatti to confirm progress and plan each stage. Worse was that Dart's and Del Grande's claim for the great antiquity of metal smelting at Mumbwa had been ruined when the laboratory results identified their 'slag' as being burnt organic material in the cave floor.

32 In preparation for his own talk at the S2A3 meeting in Durban in 1938, Smuts wrote to Drennan: 'I am anxious in my evening address at the Science Association meeting at Durban to show slides of the principal human types and skeletons found in South Africa and elsewhere. They help to keep the attention of the audience during the long strain of an address. Do you perhaps have any photos or sketches which I could use for the purpose?' Department of Human Biology, University of Cape Town, letter from Jan Christian Smuts to Matthew Drennan, 25 May 1938.

33 https://antiquarianauctions.com/lots/fine-smuts-signed-typed-letter-testimonial-re-raymond-dart, accessed 2 July 2021.

34 Lockhart also had fewer publications, but most were on traditional anatomical topics, which may have been a factor that the committee considered important. A list of Lockhart's publications was provided by the University of Aberdeen, 18 March 2020.

35 'Research Resources in Medical History', University of Aberdeen, Aberdeen, Scotland, https://www.abdn.ac.uk/special-collections/research-resources-in-medical-history-397.php, accessed 2 July 2021.

36 The announcement about Dart joining the board of Southern Cross appeared in Stimulus in February 1934: 6–7.

37 Interview with Hertha de Villiers, 7 February 1990, Parkhurst, Johannesburg.

38 The data sheets were labelled 'Expedition to Central Africa amongst Pygmies and Gorilla' because rather than set up his own data sheets for the expeditions, Wells used those that had been printed by Galloway in preparation for the abortive Italian expedition to East Africa.

39 This is an important legacy that started with Dart and continued with his students and their students. In the American system of anthropology, this broader definition of physical anthropology would intersect with social anthropology, linguistics and archaeology, but it was entirely isolated in anatomy under the adopted British system operating in Johannesburg and Cape Town (Morris 2012, S155) and the addition of the multiple focuses was an attempt to broaden a purely anatomical perspective.

40 Interview with Alun R. Hughes, 8 February 1990, Department of Anatomy, University of the Witwatersrand, Johannesburg.

41 Interview with Ralph Kirsch, 5 May 1992, Department of Anatomy, University of Cape Town.

42 Interview, Hughes.

43 Department of Human Biology, University of Cape Town, Wells unpublished papers and seminars: 'Our Present Knowledge of the Dark Age Population of South-Eastern Scotland'; 'Human Types in Prehistoric Britain'; 'Pre-Christian Burial in North-Western Europe'; 'A Review of the Physical Anthropology of Lothian'; 'A Review of the Physical Anthropology of Lothian with Special Reference to the Survival of the Bronze Age Type'; 'Skeletons from Scotland: Tables of Data'; 'The Graves of Our Forefathers: Prehistoric Burials in Western Europe'; and 'Cranial Length and Breadth in Late Palaeolithic and Epipalaeolithic Skulls'.

44 Interview, Kirsch; interview with Ronald Singer, 20 November 1991, University of Chicago.

45 University of the Witwatersrand Archives, Johannesburg, letter from Lawrence Wells to Raymond Dart, 10 August 1946.

46 Department of Human Biology, University of Cape Town, L.H. Wells, "The Skull of the South African Negro'.

47 Interview, Singer.

48 In a letter to James Wilson of 4 October 1938, Dart said that the paper would put 'a lot of old information in a new way & should, I feel, throw considerable light on the general trend of human events in Europe & Africa in the period since the last Ice Age'. Ditsong National Museum of Natural History, Pretoria, Broom archives.

49 Sergio Sergi produced several papers classifying Europeans into racial subdivisions based on his system of cranial shapes. Dart referred to several of Sergi's papers in his 1939 paper, but he also supervised Jock Gear's paper for the S2A3 in 1929 (Gear 1929) on cranial forms in native South African races, in which Gear referenced one of Sergi's (1912) publications, *Crania Habessinica: Contributo all'Antropologia dell'Africa Orientale*.

50 University of the Witwatersrand Archives, Johannesburg, letter from Peter Arnold to Raymond Dart, 4 January 1961.

51   University of the Witwatersrand Archives, Johannesburg, letter from Helen Suzman to Raymond Dart, 9 April 1966.
52   Interview, Hughes.
53   Kuljian has had a very personal look at the people who were studied by Dart in the Kalahari and in Johannesburg. She introduces us to them by name, especially !Guice, the patriarch of the group, his daughter /Khanako and his granddaughter /Keri-/Keri. Kuljian (2016) follows the sad story of /Keri-/Keri, who died in the hospital in Oudtshoorn in 1939 and whose remains were transported to Johannesburg to serve as anatomical specimens.
54   University of the Witwatersrand Archives, Johannesburg, letter from Reginald Ruggles Gates to Raymond Dart, 30 April 1946.
55   University of the Witwatersrand Archives, Johannesburg, letter from Gates to Dart, 27 December 1948.
56   University of the Witwatersrand Archives, Johannesburg, letter from Gates to Dart, 27 December 1948.
57   University of the Witwatersrand Archives, Johannesburg, letter from Dart to Gates, 27 January 1949.

6

Ronald Singer, Phillip Tobias and the 'New Physical Anthropology'

n January 1992, Frank Spencer wrote to me at the University of Cape Town to ask if I would contribute a paper to his upcoming *Encyclopaedia of the History of Physical Anthropology*.[1] This was the start of a long correspondence and collaboration over three years, which resulted in an entry on South Africa for the encyclopaedia (Morris and Tobias 1997).[2] What started as a simple request for a historical overview became a much more complicated process when Spencer realised that he had inadvertently asked both me and Phillip Tobias to contribute on the same topic separately. In a diplomatic masterpiece, Spencer managed to smooth ruffled feathers and to get the two of us to provide non-competing portions of the same entry: palaeoanthropology and primatology for Tobias, and human variation and skeletal biology for me. The two halves were then forged into one through Spencer's editorial expertise, but not without substantial friction, rooted in Tobias's desire to direct the way his own role would be seen in the history of physical anthropology in South Africa.

The central issue in the construction of the entry was my contention that Ronald Singer and Tobias had an equal role in the launch of the 'new physical anthropology' in South Africa, and that Tobias's break with typology was delayed and occurred after Singer's important Boskop paper in 1958.[3] Spencer had to use his negotiating skills to find common ground. The final published version downplayed Singer's role to satisfy Tobias, but the debate between Tobias and me was not resolved. In a letter to Spencer, I noted, 'I am also learning a great deal about PVT [Phillip Valentine Tobias] from our jousting. I didn't realise how defensive he is about his role in local anthropological history.'[4] The original pre-edit version of the paper was eventually published separately (Morris 2005b), but the friction over these issues continued. Tobias grudgingly supported my application for a Fulbright scholarship to the United States in 1995, but wrote to me with specific criticisms of the proposal with respect to the history of physical anthropology.[5] Among the issues raised were Singer's own self-aggrandising opinions about himself; the need to study the Afrikaans literature, with the implication that the roots of apartheid must be traced there and not in the English literature; the accusation that I equated typology with racism; and the argument that the social context of South African English scientists was less important than the rise of Afrikaner nationalism. My long response to Tobias's letter refuted each of these points, but also noted how Tobias was positioning himself as the authority of historical interpretation.[6]

These events of some 25 years ago highlight how being too close to historical events can colour one's view about them. Much of Tobias's concern stemmed from his own venture into the history of physical anthropology with his 1985 paper (Tobias 1985a). Tobias's paper remains the most thorough accounting of the entire field in South Africa and its details of names and projects are unparalleled, but the paper makes no real effort to examine the context and motivations of the long list of scientists he recounts. This was the heart (and heat) of the debate between Tobias and me, a debate that still needs serious consideration.

PHILLIP TOBIAS: THE EARLY YEARS

In his book of taped interviews, Tobias reveals that he found it difficult to write about himself (Tobias, Štrkalj and Dugard 2008, 8). To some extent this is true, as Tobias struggled to produce the first volume of his autobiography (Tobias 2000, 36). After eventually publishing *Into the Past* (Tobias 2005), he planned a second volume, but did not complete it. Despite that, his life publication list exceeded 1 300, a feat very few academics can match. In addition, there are numerous published interviews, transcriptions of public lectures and films, and seventeen years' worth of letters of seasonal greetings, some of which run to ten pages or more. He also kept personal journals, which detailed his activities, including names and dates of people whom he met, which he used to jog his memory as he wrote his autobiography. There is also a wealth of personal letters in various hands as Tobias was a correspondent of note, trading letters with the wide range of individuals who made up his long list of acquaintances. Biographers are not going to struggle to find material on Tobias, but they will struggle to edit the vast volume into a manageable amount. A detailed account of Tobias's life is beyond the scope of this work, so other than a brief account drawn primarily from his autobiography (Tobias 2005), the focus will be on two critical aspects of his life that have a direct impact on the nature of his research: his political awakening and his broad approach to anthropological research that developed during his career.

Phillip Tobias (1925–2012) was born in Durban, the son of Jewish immigrants from England. His father owned a downtown toy shop, but the business was not a success and after it failed his parents split up, with his father going to Johannesburg and his mother moving to Bloemfontein with the children to live with her family. Tobias did not have happy memories of his childhood. When he completed junior school, the family decided that he should return to Durban with his mother to complete high school. His only sibling, his sister Valerie, died of diabetes while Tobias was still in high school in 1942. He had already decided by this time that he would study medicine to see if he could understand the genetic reasons why his sister and his aunt had died of diabetes, but his mother had not shown any sign of the disease. Tobias attributes his occasional visits to the displays at the Durban Natural History Museum for his interest in archaeology, the mechanisms of genetics and animal biodiversity.

Although one of his close friends from Durban High School tried to get him to study medicine with him in Cape Town, Tobias chose to apply to the University of the Witwatersrand (Wits) where he could be close to his father in Johannesburg. He began the first year of the MBChB (Bachelor of Medicine, Bachelor of Surgery) programme in 1943. In his autobiography Tobias notes how he was smitten by science right from the start of the first-year programme's lectures on embryology from the zoologists and ecology and genetics from the Department of Botany, but it was his introduction to anatomy in his second year that provided his future scientific direction.[7] Tobias mentions how both Christine Gilbert and Joseph Gillman cemented his passion for embryology and histology, but it was the 'unforgettable, exceptional personality' of Raymond Dart that drew him into the structure and function of the human body and eventually into the study of physical anthropology (Tobias 2005, 24).

By his own admission, Tobias did not have political views when he arrived at Wits in 1943. He embraced student life with a passion, but he was especially conscious of the multi-ethnic dimensions of the Wits undergraduate population because he had come from a high school at which the students were overwhelmingly white. Officially, the university had a policy of non-segregation and students of colour were accepted into the academic programme, but socialising between groups was discouraged. Tobias, along with Sydney Brenner and Harry Seftel, were three of only a few white medical students who actively socialised with all of their fellow students. Tobias and Brenner were part of a growing student intelligentsia at Wits (Murray 1997, 74). Tobias was part of the left/liberal group (Murray 1997, 103). Both Brenner and Tobias were wonderful orators and Brenner became involved in the Students' Representative Council (SRC) while Tobias joined the National Union of South African Students (Nusas). Tobias marks the start of his interest in politics to the events of 1945 when Nusas invited the University of Fort Hare (a university for black students in the Eastern Cape) to join. This initiated a series of student debates and caused a split between the Afrikaans- and English-medium universities. Tobias took an active role in Nusas at this point and was elected president of the student organisation in July 1948.

Tobias became president of an organisation that was disintegrating. The Fort Hare issue had already caused all the Afrikaans-medium schools to withdraw from the organisation in order to set up their own union, which would embrace 'white Christian National ideology'.[8] But the split ran deeper, with many of the English-medium universities also threatening to withdraw. The issue for them was the position taken by Nusas to withdraw the social colour bar. Tobias's own position was in support of a complete social and academic non-racialism, but he had to find a course that would maintain his own ideals yet allow the more conservative view as well. In order to do this, he shifted the debate away from the social questions to the broader issue of state interference in university autonomy (Murray 1997, 120). During his three-year term as president of Nusas, he managed to unite the four liberal universities (Wits, the University of Cape Town, Natal and Rhodes). The withdrawal of the study permit for Eduardo Mondlane of Mozambique and the denial of permission for

Indian students from Natal to study at Wits were central issues about academic freedom on which Tobias and Nusas would take a strong stand. In his 1950 Nusas presidential address, Tobias warned that the changes in government policy that stopped government bursaries to black students at Wits was a 'prelude to legislating apartheid for the universities' (Murray 1997, 114). Under Tobias, Nusas launched the first anti-apartheid campaign in the universities, which garnered substantial student support. The campaign would not get the government to change its mind about bursaries for black students, but it did force Wits to provide its own funding for these scholars (Tobias 2005, 60). The campaign continued after Tobias had completed his term as president. In February 1954, a university convocation led by Tobias voted 180 to 6 to officially declare Wits's opposition to university apartheid. The motion was sent to senate where the whole issue of academic and social integration was hotly debated, with many of the academic staff taking the position that they did not want black students in the university at all.[9] The motion supporting the convocation was eventually passed by a vote of 24 to 10.

Tobias's active role in student politics drained what was left of his spare time in a very busy schedule, which did not include much of a social life. He slept very little as a student because he needed to work to pay for his university fees. During the day he was employed as a demonstrator in the Department of Anatomy, but at night and in the evenings he studied and wrote magazine articles, which paid quite well. He had four lives as a student: studying for his medical degree; working on his Master's/doctorate; earning money; and playing his role in student leadership (Tobias, Štrkalj and Dugard 2008, 7). His presidency of Nusas was officially part-time, but Tobias was so busy that Nusas made the position full-time after he left. Although he remained sympathetic to the ideals of Nusas throughout his life, he chose to have no party affiliations, other than a brief membership of the Liberal Party when it was under Alan Paton.

The year 1945 was not only Tobias's introduction into politics, it was also his introduction into the world of archaeology and fossils. The science students in the department were given lectures on fossil hominids by none other than Robert Broom and then went on a field excursion in May to the fossil site at Sterkfontein (Tobias 1997, 72).

Tobias found the visit to Sterkfontein so interesting that he and fellow student Joseph Stokes Jenson developed the idea of leading a student expedition to a similar place in the July vacation. It was Dart who suggested Makapansgat near Potgietersrus in the northern Transvaal (as it was then). Dart had been sent fossils from there in 1924, but he suggested that Tobias and Jenson go to see Professor Clarence van Riet Lowe, as he had worked on both Early Stone Age and Middle Stone Age tools in the exposed deposits at the Cave of Hearths in the Valley (Tobias, Štrkalj and Dugard 2008, 40). Van Riet Lowe was very supportive, and not only did he give them access to his 1930s' field files, but he also put them in touch with the Maguire family of Potgietersrus, who would guide them into the site.

The two students led a multidisciplinary expedition to the valley in June and July 1945, and Tobias led a further five expeditions to the valley between then and 1947. He missed the July 1946 expedition as he used his student vacation to join Van Riet Lowe and Berry Malan at their excavation of Rose Cottage Cave in the Free State. This was Tobias's opportunity to learn proper archaeological field techniques (Tobias, Štrkalj and Dugard 2008, 45) and directly led to his excavation of the site at Mwulu's Cave in the Makapansgat Valley in January 1947. Considering that Tobias was only twenty years old in 1945, this demonstrates both his determination and his competence on many fronts. The experience was formative, launching him into the fields of archaeology, geology and palaeontology, but it also had a critical consequence for Dart. After his failure to convince the scientific world of the importance of his discovery at Taung, Dart had retreated from fossil studies. Tobias returned from his first expedition with fossils

Figure 6.1: Robert Broom at the Sterkfontein fossil site in August 1936. He is indicating the exact place of discovery of the specimen TM1511, which was the first adult australopithecine fossil discovered. The man to his left is G.W. Barlow, the quarry foreman, and in the background are three staff members from the Transvaal Museum. (Source: Findlay 1972)

collected from the lime miners' rubble at the limeworks site – the same site that had provided specimens for Dart in 1924. Among the specimens was the skull of a fossil baboon, which Dart recognised as the same species linked to the australopithecine site of Sterkfontein. Dart's interest in fossils was rekindled and between 1947 and 1966, he oversaw new systematic excavations at Makapansgat that produced a wealth of australopithecine fossils (Tobias 1997, 78).

Tobias completed his MBChB degree at the end of 1950, but chose not to complete the houseman year in order to be registered by the South African Medical and Dental Council as a medical practitioner in South Africa. Lawrence Wells had resigned to take up a post as physical anthropologist at the University of Edinburgh and Dart offered Tobias the position that Wells had vacated as a lecturer. This choice placed Tobias on a career trajectory to academia rather than to clinical practice and he never completed the intern year.

Tobias had barely begun his full-time academic duties when Dart asked him to join an expedition to the Kalahari. The Panhard-Capricorn Expedition, led by the Frenchman François Balsan, planned to cross Africa from west to east, roughly along the line of the Tropic of Capricorn. The specific objectives after departing from Johannesburg were to attempt to find the Lost City in the southern Kalahari, as described by Gilarmi Farini in 1886 (Farini 1973, 357), then to travel west to Walvis Bay in South West Africa (Namibia) to begin the west to east transect. En route they planned to study the language and songs of the San of the Kalahari and visit the Tsolido Hills to find rumoured rock paintings. The final leg would be through Serowe in Botswana, then through the northern Transvaal, meeting the Indian Ocean either at Vilanculos or Inhambane before returning via Lourenço Marques to Johannesburg (Balsan 1954, 7–8). The Panhard Automobile Company had provided two ten-ton trucks and there were six French members of the expedition plus two drivers. The expedition approached the South African government for help in obtaining a suitably qualified scientific member, and the government in turn approached Wits. Van Riet Lowe and Dart recommended Tobias.

Dart made it very clear to Tobias that his highest priority should be the study of the 'Bushmen' (Tobias, Štrkalj and Dugard 2008, 87), but Tobias had not done any anthropometry and he needed to teach himself how to take measurements. Tobias had two further attributes: he was medically qualified (although not registered) and he could speak Afrikaans. The expedition lasted from August to October 1951 and was as much an adventure for Tobias as it was a scientific mission. His abilities in Afrikaans were used frequently, as many of the people they visited used that tongue for communication with outsiders. He also provided medical care for sprains and cuts, but fortunately nothing more serious than that. His own health was problematic as he suffered from asthma and found that he had equine allergies, which were an issue since donkeys were common as the main form of local transport. Tobias also became the de facto archaeologist on the trip because the French expert, Jacques Mauduit, specialised in European art and pottery and could not even recognise African Later Stone Age and Middle Stone Age tools (Tobias, Štrkalj and Dugard 2008, 87). Tobias's hunting for implements provided Balsan with a great deal of amusement. Tobias took every opportunity to look for archaeological specimens and was very excited when he found relics. He collected nearly ten kilogrammes of stone tools while he was stranded on the Nossop waiting for spare parts for the one truck, and he was convinced that the Nossop Valley was especially important archaeologically. Tobias was so keen that he searched for artefacts even when everyone else was relaxing after a busy day. One evening at Rietfontein, while everyone else tried to banish the cold beside the large campfire, Tobias was 'stealing about like a thief, stooping now and then to pick up an interesting stone by the light of his torch' (Balsan 1954, 105). It was Tobias who convinced Balsan that Farini must have misinterpreted the flat eroding shelves of limestone as ruins and that no ruins actually existed at the Nossop location.

Figure 6.2: The members of the 1951 Panhard-Capricorn Expedition photographed just before their departure for the Kalahari. Phillip Tobias, wearing a beret, is in the centre of the picture. To his left is the expedition leader, François Balsan. (Source: Balsan 1954)

But the most important aspect of the Panhard-Capricorn Expedition for Tobias was the opportunity to gather data on the San. Reporting back to Balsan after their vehicles had been separated for several weeks, Tobias said, 'I managed to get eight more facial moulds, and they all succeeded except one – a woman who thought she was suffocating under the plaster so she tore it off and ran away. And then, along with the moulds, I've collected about fifty sheets of measurements and innumerable medical and mental reports' (Balsan 1954, 193–194). Balsan was fascinated by Tobias's face-moulding process. This was the procedure originally shown to Dart by Lidio Cipriani and adapted by Dart's senior technician Eric Williams. According to Balsan, Tobias

> worked wonders, as much with his tongue as with his fingers. He contrived to mix the plaster to the right consistency while at the same time smearing on the Vaseline. In his gentle, soothing voice he murmured sweetly – if incomprehensibly – to his model, charming him into co-operation. When the time came to put the tubes up his nostrils, he was convinced that Doc was doing him a great honour; the white daub seemed a cool and kindly protection against the sun; and still my dear Tobias kept on talking, talking … until the heat had dried out the plaster into a shell, and his work was finished. (Balsan 1954, 93)

Tobias published eight papers using the data gathered in 1951 on the Auen (=Au//ei) and Naron (Nharo) San from the Ghanzi area and an additional three papers extending these data into the broader topic of the evolution and population size of these people. The first paper (Tobias 1954) presented the data for a racially hybrid family, which was in response to the debate over the United Nations Educational, Scientific and Cultural Organization's (Unesco's) *Statement on Race* (Montagu 1951), in which critics had said that 'Bushmen' and 'Pygmies' were so different from other forms of humanity that they could not interbreed with them. Although the objective of Tobias's paper was to refute such genetic nonsense, his paper reads like much of the earlier literature that concerned itself with the inheritance of racial characters. Although Tobias's work was based on genetics, there was little real understanding of how the physical characters were inherited. The four papers in French (Tobias 1955/56), along with a separate paper in English (Tobias 1957), were the main results of the research and covered cultural and demographic aspects, through non-metrical classical traits to metric analysis and then a final paper on general conclusions. Much of this final paper referred to the various hypothetical types that Dart and the other typologists had created.

In his autobiography and in his taped interviews, Tobias commented on how the experience of meeting the San shaped his understanding of how human populations adapted to their environment, but this was not completely evident at the time. Looking back at these papers, he remarked, 'I realise now that I was under the influence of Raymond Dart's typological approach to the analysis of the "racial affinities" of African peoples. It was an influence that I was soon to shake off, for it was an approach that was totally at variance with the lessons of human genetics' (Tobias 2005, 68).

In the early 1950s Tobias was still very much on the path to specialising in human genetics. He had begun work on the chromosomes of the laboratory rat in his science year and followed this with other rodent species for his doctorate. He still had in mind the idea of specialising in human genetics although there was no one in the department who had the knowledge and expertise to guide him. He was still vacillating about this when he received news that he had been awarded a Nuffield Senior Fellowship to go to the United Kingdom for a year from January 1955 and this might be followed by a Rockefeller Travelling Fellowship to visit academic departments in the United States if this second application was successful. In total he would be away from South Africa for eighteen months.

Tobias chose the Department of Archaeology and Anthropology at the University of Cambridge as his base. This was the location of the Duckworth laboratory of human skeletons, headed by Professor J.C. (Jack) Trevor. He was one of seven physical anthropologists who had been invited to contribute to the second Unesco Statement on Race, along with Solly Zuckerman and the population geneticist J.B.S. Haldane (Tobias 2005). Trevor was also a leading figure in the biometric school of anthropological research, which focused on mathematical or statistical approaches to anthropology, an area in which Tobias had no knowledge.

The biometric school had developed under Karl Pearson in the United Kingdom at the start of the twentieth century and was a methodology in opposition to typology. Pearson had argued that in order to understand evolution one must study the very deviations that the typologists ignored (Stepan 1982, 120). It was Pearson's intent to use biometrics to replace typology, not racial classification itself. In fact, Pearson was closely allied to the eugenics movement and saw biometrics as applied anthropology to be used to study 'racial problems in man' (Pearson 1925, 1). Pearson openly used a large assortment of different social, religious and other groups as 'races'. He solved the problem of definition of groups by not defining them. Biometric analysis was not popular among the South African typologists and Dart did not use statistics at all. His analyses relied on the visual representation of features, something that influenced Tobias strongly in these early years. Looking back on his time in Cambridge in his autobiography, Tobias quotes from his own journal, saying that the biometric school had a preoccupation with methodology and that 'measurement and statistical analysis have become a substitute for *thinking*' (Tobias 2005, 78). When Tobias agreed to be internal examiner for the Master's thesis of Hertha de Villiers in the following year, he expressed concern about the statistical methods that De Villiers planned to employ. He wrote to Dart, suggesting that the project might be 'enough without the biometrical analysis'.[10] De Villiers would eventually complete a doctorate with Tobias as supervisor that would include the statistical analyses that Tobias worried about. While he was in Cambridge, Trevor suggested that he look at the curvature of the occipital bone on the specimens in the Duckworth collection using basic measurement techniques. Although Tobias would change his mind about typology in future years, he never did come to use complex statistics in his own research.

Tobias loved the Cambridge academic and social environment and referred to his stay there as 'the happiest year of my life' (Tobias 2005, 70). But Cambridge was also a base from which he could visit other centres of anthropology in Europe. During his year of residence, he visited the anatomist Wilfrid Le Gros Clark in Oxford. He told Le Gros Clark of his dual interests in cytogenetics and anthropology, and it was Le Gros Clark who warned him that he would struggle to do science on two fronts. He had to choose one or the other (Tobias 2005, 94). Also, while in Cambridge, Tobias was a visitor to the Natural History Museum in London where he took the opportunity to see the Broken Hill (Kabwe) skull and some of the specimens that Louis Leakey had stored there. One of Leakey's specimens was an archaic-looking mandible from the site of Kanam in Kenya. The context was inferred to be very ancient, but the mandible had a chin, which was a modern feature. Tobias analysed the specimen both morphologically and microscopically and found that the chin was in fact a pathological growth after an injury. It was working on this specimen that finally tilted Tobias's compass toward anthropology rather than genetics (Tobias, Štrkalj and Dugard 2008, 100).

Tobias was notified that he had been awarded a Rockefeller Travelling Fellowship just as his Nuffield Fellowship was ending. The trip to the United States was going to be very different from his sojourn in Cambridge. He would not be based in one location and would ultimately travel to more than 30 university departments, museums and research institutions over a period of six months. Tobias himself considered this trip so important that he devoted 8 chapters and 58 pages of his autobiography to the events that transpired there. The people and places he visited embraced the whole range of his interests, from anatomy to human genetics and of course physical anthropology. There is no need to go into detail about his activities, as these are well covered in his own writings, but a few of the people he met would have a significant influence on his future career. In Ann Arbor, Michigan, Tobias visited James V. Neel in the human genetics department. He was very impressed with the fieldwork on the genetics of hunter-gatherers that Neel was doing in Brazil and this was to lead directly to Tobias hiring Trefor Jenkins as a lecturer in human genetics in 1963 before Wits set up its own Department of Human Genetics. In Chicago, Tobias met Sherwood (Sherry) Washburn, who was launching his own vision of a new physical anthropology. Tobias was critical of him not because of any specific research, but because Washburn did not include enough data from non-American sources and was failing to teach his students the history of the subject. Washburn tended to ignore earlier research as if he was trying to make a clean break with the past. The problem was that Washburn was training a very large cohort of students who would not be aware of earlier research. Tobias later says, 'It was this aspect of the "New Physical Anthropology" that gave me the most cause for alarm' (Tobias 2005, 139). In Washington, Tobias met T. Dale Stewart, who had taken over the osteological research at the Smithsonian Institution. At that time Stewart was working on the Korean War Project, which would be one of the most important early works in forensic anthropology. In Philadelphia, Tobias met Carleton Coon who was still at the peak of his career before the publication of his problematic *Origin of Races*, and J. Lawrence Angel, who to a large extent opened up the study of palaeopathology in the United States. In Cambridge, Massachusetts, Tobias was hosted by William (Bill) Howells who provided him with an introduction to the scientific high and mighty of Boston.[11] Howells hosted a dinner party for Tobias at which he was introduced to Alfred Romer (comparative palaeontologist), William C. Boyd (immunologist working on blood groups and races), Hallum Movius (Palaeolithic archaeologist) and John O. Brew (archaeologist and director of the

Peabody Museum).[12] Howells also introduced Tobias to Laurence and Lorna Marshall. Marshall had taken his family on a series of research trips to Bechuanaland (as Botswana was then known) to study the culture of the San. The Marshall trips were unrelated to the Panhard-Capricorn Expedition, but Tobias was very interested to share his experience with the Marshalls. Tobias was actively planning his own new expedition to the Kalahari under the planned Wits Kalahari Research Committee (KRC). He had already written to Joseph Weiner in Oxford suggesting that Weiner might like to join them and work on the ecological aspects of the San, although the committee did not meet until Tobias's return late in 1956.[13] The first Wits expedition to the Kalahari took place between August and September 1958, with Weiner as part of the multidisciplinary contingent (Tobias 1975, 75).

RONALD SINGER: THE VIEW FROM CAPE TOWN

Ronald Singer (1924–2006), born in Cape Town, was a year older than Phillip Tobias. Like Tobias, his family background was Jewish, but where the Tobias family were anglicised German Jews, Singer's parents were first-generation Lithuanian immigrants. His father, Solomon, had arrived in South Africa just after the South African War and was set up as a *smous* (travelling salesman) by the Jewish Relief and Resettlement Committee. He met Ronald's mother in South Africa, and after they were married in 1910, the Singer family settled on a dairy farm just outside the small town of Malmesbury, about 70 kilometres north-east of Cape Town in the farming district known as the Swartland. Singer spent his boyhood on the farm and in Malmesbury, but his father sold the farm and moved to Muizenberg in Cape Town when Ronald was fourteen.[14]

Both Ronald and his older brother Martin studied medicine at the University of Cape Town (UCT), Martin graduating with honours in 1944 and his brother following him in 1947. Singer completed his houseman year in Kimberley in 1948 and then continued with a second year in clinical medicine at Gwelo in Southern Rhodesia (now Gweru, Zimbabwe). He remained registered with the South African Medical and Dental Council until 1973, but he did not hold a clinical post after 1949. As a student, from 1943 he worked as a demonstrator and prosector in anatomy and upon his return from Rhodesia in 1949 he was offered the post of lecturer in anatomy under Matthew Drennan. He was promoted to senior lecturer in 1953 and associate professor in 1960. Singer had a keen interest in gross and skeletal anatomy and had very much impressed Drennan with his interest and skill as an anatomist. Singer requested permission to study the skulls in the Department of Anatomy at Wits in 1950 and Drennan wrote a formal letter of introduction to Dart in support of Singer's request:

> I think he intends becoming an anatomist. At first I thought he was only a 'passing ship', but he has turned out to be a serious and hardworking student of Anatomy. He was recommended by Dr Keen to act for him in his absence and for the last six months Dr Singer and I have been 'holding the fort' whilst Dr Keen was on leave. I have found him a tower of strength and very cooperative and helpful in every direction.[15]

The Department of Anatomy at UCT was under stress at the end of the Second World War. The number of students had increased to nearly 300 as returning servicemen and women began to study medicine and Singer was one of the bright students hired to help in this crisis. De Villiers was a student there between 1943 and 1945 and remembered Singer as a demonstrator. There was a separate dissection hall for women, with twelve bodies available, but the classes were enormous and they had to be taught in double shifts. There was a shortage of teaching staff, resulting in no extra tutorials and mentoring and a high failure rate.[16]

Singer had not entered the medical programme with any specific research interest in mind. His time was occupied with his own medical studies along with the departmental demonstrating and, as he reached the latter years of the course, with private coaching for surgical students preparing for their primary examination. What did awaken his interest was the collection of archaeological skeletons accumulating in the departmental museum. His conversations with Drennan between classes often covered anthropology and he spoke to Jack Keen about his analysis of the Peers Cave specimens from Fish Hoek (Keen 1941). Singer compiled a literature survey about the 'Bushmen' in 1945 and submitted it for the Cornwall and York Prize. His submission was unsuccessful, but he did publish the work in the UCT medical students' journal, *Retina* (Singer 1946). Singer's conclusions showed how little first-hand knowledge he had of either the skeletons or of Khoesan people in real life. He had no reference point to judge the earlier research, so simply followed old conclusions. He noted how the 'Bushman' anatomy was towards the simian end of the human scale, which he said confirmed that 'the Bushman is undoubtedly a member of one of the lowest of human races'. Comparing the San to the European, he said, 'We see the one crowned with all intellectual and spiritual glory of the race, while the other still occupies the lowest scale in human existence' (Singer 1946, 4). This was hardly a prestigious entry into the field of anthropology, but when he returned to UCT's Department of Anatomy as a lecturer he ventured once again into the field with another submission to the Cornwall and York Prize. He focused on the discovery of the fossil and sub-fossil specimens in an essay titled 'Discoveries in Southern Africa Contributing to Our Knowledge of the Ancestry of Man' (Singer 1950). This time, the committee was more impressed and awarded him the prize, along with a purse of £60.

Singer's Cornwall and York essay was very different from his previous effort. He had been busy teaching himself palaeontology, along with embryology and human genetics. He would have lunch regularly with Drennan in Drennan's office anteroom and through him learned about the scientists in the field (although little about anthropology itself). He began a study of growth in the human cranium in order to understand cranial variation. It was the Cornwall and York paper that got him thinking about the Boskop skull, and how poorly the Boskop type was defined, and about fossils in general.[17] His interests were further aroused when Drennan told him that a local farmer had recently brought in some fossil animal bones. Drennan had recommended that the farmer give them to the South African Museum, but Singer decided to follow up on the discovery, which led directly to his exploring the site at Elandsfontein in May 1951 (Singer 1993, 105).

In the same year that Singer won the Cornwall and York Prize, he was told that his application for a Rotary Foundation Fellowship to the United States was successful. Although still fascinated by anthropology, his real interests lay in experimental embryology and he had already applied for a placement at the Department of Embryology at the Carnegie laboratories in Baltimore in anticipation that he would win the Rotary fellowship. Singer left for the United States in the middle of 1951 and would be gone for fourteen months. His official placement was jointly in the departments of biology and embryology at the Carnegie laboratories at Johns Hopkins University, where he spent thirteen months learning the intricate laboratory methods of experimental embryology. Singer began a project on osteogenesis, with plans to continue when he returned home to Cape Town. One of the conditions of the fellowship was that he had to give talks to other Rotary clubs across the country, and he took this opportunity to visit twelve of the largest anatomy and anthropology departments in the United States. Although his formal time was spent working in embryology, his interest in anthropology continued. He used the close proximity of Baltimore to Washington to do research at the Smithsonian

Institution, where he received special permission to work in the skeleton collection over weekends:

> I got permission from the Smithsonian from Stewart, who was then Head of the Anthropology section, I got keys and permission, which was very rarely given, to work there at night and on Sundays and Saturdays, when the Smithsonian was closed. So I actually spent every weekend studying skulls and skeletons at the Smithsonian Institution … I was working at that time on variation of closures of the sutures of the skull and variation in growth of the sutures. I measured every single part of every suture that was known according to the various textbooks and the closure and the size of the left [and right] and I was going to come up with some grandiose scheme.[18]

Singer added the Smithsonian measurements to the data he had already gathered in Johannesburg and published a paper that showed that the sequence of fusion of the cranial sutures was not constant and that it was an imperfect method of estimating age at death (Singer 1953a, 58).

Working with the skeletons in Washington and his discussions with anthropologists in other departments reinforced his interest in skeletal variation, especially in relation to the fossil material he had discussed in his Cornwall and York paper. Before returning to South Africa, Singer spent a month in England at the British Museum (Natural History) and the Royal College of Surgeons in London, and at the Department of Anatomy in Oxford. He examined the ancient and modern skeletons in these collections while he pondered the definition of *Homo sapiens* and how it differed from Neanderthals. He also used the specimens in the collections to look again at the differences between 'Bushmen' and 'Hottentots' and how the Boskop type confused the issue. This was when he first began to realise that these types were being defined by physical descriptions that were entirely subjective.[19]

Singer planned on setting up a research laboratory in experimental embryology on his return to Cape Town, but his plans were dashed by the departmental head. Drennan absolutely refused to provide space and funding and insisted that Singer continue in anthropology, where he felt Singer's strength lay. Singer was furious, but could do nothing to change Drennan's mind. He continued with his exploration of the fossil site at Elandsfontein and began to collaborate on anthropological topics with his colleagues in the Department of Physiology. Singer wrote a literature survey on the sickle cell trait in Africa, using information from the literature and data provided by his physiology colleague, Dr Otto Budtz-Olsen (Singer 1953b, 645).

Research questions raised by this work so intrigued Singer and Budtz-Olsen that they spent six weeks in mid-1954 gathering blood data on Malagasy populations throughout Madagascar, funded by the UCT Faculty of Medicine and the Wenner-Gren Foundation for Anthropological Research. They sampled more than 600 people (Singer et al. 1957, 98). The UCT Department of Medicine and the Council for Scientific and Industrial Research (CSIR) Institute of Clinical Nutrition had already done their own clinical expedition to assess the general health of the San in northern South West Africa in 1952, and Budtz-Olen had drawn blood samples from more than 1 300 !Kung and Heikum San. This had given Singer a substantial comparative sample for his Malagasy analysis, but he realised that even more data would be required for a serious discussion of blood group variation. He joined forces with Weiner of the University of Oxford to extend these data, first during 1958 among the Nama in the Richtersveld on the South West African border, and then in South West Africa in July and August 1961 when they studied the Bondelswart Khoekhoe near Keetmanshoop and then Cape Coloureds and Rehoboth Basters in the Rehoboth area just south of Windhoek.

Singer now found himself balancing research topics in palaeoanthropology, palaeontology, human biology and anatomy. The discovery of the Saldanha ancient human skull from the fossil site at Elandsfontein in 1953 put him on the international map of palaeoanthropology. He managed to secure funding to visit museums in southern Africa and in 1956 and 1957 received leave from UCT to continue this research in museums in Europe. Singer attended the fifth International Congress on Anthropology and Ethnology in Philadelphia in September 1956 and presented a paper criticising the Boskop concept. Dart was in the audience and he launched a personal attack against Singer at the end of the paper. Dart accused Singer of destroying science in South Africa by 'belittling the work of science that has been done there'.[20] This permanently damaged relations between Dart and Singer, but did not stop Singer from publishing the paper that would effectively destroy the validity of Boskop as a racial type (Singer 1958).

When Drennan retired from his chair in Cape Town at the end of 1955, Singer was not yet ready to apply for such a senior position, especially when he heard that Wells had expressed an interest in the post. The appointment of Wells meant that a headship in Cape Town would not be available again for some time. When the University of Stellenbosch began to set up its own medical school in 1956, the dean at Stellenbosch approached the newly retired Drennan to ask if he felt Singer would make a good appointee. Not only did Drennan support Singer, but he forwarded his curriculum vitae to Stellenbosch. Singer was uncertain how to take this. The faculty at the University of Stellenbosch was staunchly Afrikaner nationalist and Singer worried that he would not fit in. Drennan recommended that, if he was appointed, Singer should just focus on teaching and research and leave the politics alone. The search committee indeed selected Singer and his name was passed on to the university senate for confirmation. But the senate would not approve Singer and instead recommended

Johan Ferdinand van Eyck Kirsten, an Afrikaans-speaking zoologist. At this stage Singer was not aware of his selection by the committee, but 'somehow or other, some member of the Cape Town Council, a man called Berman, found out that I had been turned down in favour of Kirsten, who hadn't applied at the time I was appointed. He [Berman] then felt this was anti-Semitic and pro-Nazi.'[21] Singer found out about this from the pages of the local newspapers and although he was disappointed, he was fatalistic about the situation. However, he felt differently about the Wits chair when it became available on the retirement of Dart at the end of 1958. He applied for that position, but suspected that the committee would prefer Tobias as the home-grown candidate. In the end he was right and the Wits committee appointed Tobias. Singer was again disappointed, but he had already made the decision that he would leave South Africa – what was driving him away was politics.

Singer had been disturbed by the social and academic segregation in the Faculty of Medicine at UCT. The number of Indian and coloured students had been slowly growing from the late 1940s, rising to some 12 per cent of the UCT student body in 1959 (Phillips 2019, 338). According to Singer, neither Drennan nor Jack Keen actively discriminated against students of colour, but they tended to ignore discrimination when it happened.[22] The senior technician in charge of cadaver allocation, 'Oom' (Uncle) Daan Coetzee, made sure that coloured students did not dissect white bodies. A survey of staff memories before 1970 noted that classes and dissection halls were racially segregated at that time (UCT 2003, 153), but this may have reflected Coetzee's actions, rather than a formal system of physical separation during dissection. Singer, Ted Keen and De Villiers did not mention separate dissection venues, although De Villiers did note that women dissected in separate venues from men.[23] Of greater impact was the exclusion of students of colour from post-mortems if the body was of a white person (Phillips 2019, 104). According to Singer, 'When there were white patients to be used as demonstrations in lectures by professors there was always a sign on the door of the lecture room, and there was a W put on the door that meant that a white patient was inside and coloureds could not come in.'[24] The effect of this was exceptionally demoralising to students of colour (Digby 2013, 277). This is what politicised Singer. The last straw was the introduction of the Extension of University Education Bill in 1959, which would segregate the universities and result in the exclusion of most of the coloured and Indian students from UCT. The promulgation of the Act was the subject of an intense debate in the UCT senate. Singer spoke at the meeting, saying that they should threaten to close the university if the Act went through: 'I am not saying we mustn't keep it open. I'm just saying we must all threaten to resign and threaten to close the university. Whether we carry out that threat or not, we must let the world know how we feel about this thing.'[25] Singer refused to sign the UCT declaration on apartheid because he thought it was hypocrisy if they did not threaten to close, but the majority were too concerned about losing their jobs and the motion before the UCT senate did not pass.

How political was Singer? Unlike Tobias, who proclaimed his opinions loudly in debate and in print, there are no political statements from Singer. His obituary in the *Anatomical Record* remarks that his 'active opposition to the policies of the apartheid regime had made life there untenable' (Dechow 2006, 115), while the *Chicago Tribune* obituary spoke about his politics clashing with government policy.[26] Singer may have exaggerated his anti-apartheid activities a bit for his American audience, as Ralph Ger who was in the Department of Anatomy at UCT with Singer noted that he was not active in apartheid resistance.[27] Ger was investigated by the security police for providing medical help to anti-apartheid activists. He kept a dossier of security prisoners who were brought into Somerset Hospital and their injuries and gave it to the Liberal Party in 1958, which resulted in his being fired from his job as surgeon at the Somerset and Groote Schuur hospitals. He went into private practice for a period after this, volunteering to teach anatomy at UCT, but without pay or formal recognition. He found that his anatomical colleagues were unwilling or uninterested in supporting his political views and, in the end, he felt he had no choice but to leave the country.[28] Singer was certainly sympathetic to the anti-apartheid cause. His wife was a member of the Black Sash and Singer was aware that one of his assistants was using the duplication machine in his office to produce political pamphlets, but he was unlikely to have been on the radar of the South African security police.[29] Certainly, at the time people in South Africa were concerned that they could be in trouble if their political views differed from the government of the day, but in practice the police only acted if one associated with known activists or if one's views had an impact in the press or in wider society. Singer was not driven out because of his political activities, but his decision to leave was caused by his disgust with the apartheid system: 'I was hoping for years and years, still doing my little bit in politics hoping that things would change, but as the years went by it [apartheid] just became irreversible.'[30]

Singer's research in anthropology was well regarded internationally and he obtained a post as visiting professor at the University of Illinois in 1959–1960. He was keen to see if his wife and children would like the United States. The University of Illinois offered him a job, but when the University of Chicago also offered him a position, he turned the Illinois post down. The Chicago position was ideal for him, but there was a catch. The normal paperwork for residence in the United States could take up to two years to process, but one option to speed this up was if he came as a political refugee. His contact at the University of Chicago applied for his refugee status and he subsequently received a letter from Robert Kennedy, the attorney general, admitting him to the United States. Singer rationalised his action: 'I was a political refugee in the sense that I was leaving my country because of politics. So, I didn't feel false at all. I wasn't in a sense in jail, my life wasn't in danger, but it would become a danger if I went on saying the things I believed in.'[31] The reality was that he was unhappy with the political situation in South Africa, his options for promotion were limited and the Chicago placement was a career advancement.

Before his arrival in Chicago, Singer had developed several lines of research that were ongoing and could be managed from the United States. His human biology research in South West Africa (Namibia) continued, working with the Oxford team of Joseph Weiner and Geoffrey Ainsworth Harrison, and later with Kunihiko Kimura of Japan. Singer's South West Africa research was not part of the International Biological Programme (IBP) set up by Weiner, but it used a similar approach to gather as wide a range of data as possible, so that the people could be understood in their anatomical, clinical and social context. Morphology and blood genetics were studied along with radiographs, fingerprints, chromosome spreads and genealogies.[32] A lot of the latter focus was on growth and the distribution of genetic abnormalities.

Singer became an honorary curator at the South African Museum when he was still at UCT Medical School as both he and the UCT committee felt that the recovered material from the excavations at Elandsfontein would be best stored at the museum. He maintained this position when he left for the United States, which meant that his research assistants could make use of laboratory space in the museum for the analysis of the excavated material (Summers 1975, 213). Excavations wound up at Elandsfontein in 1966 and Singer moved to the nearby site of Langebaanweg. In 1967 he opened excavations at Klasies River which, although more recent in time than the other two sites, were to provide a larger sample of early human remains. Singer was in a strong financial position, with funds from the American National Institutes of Health, and he was able to provide long-term employment for Peter Saunders, one of his South African excavators, and the more qualified English archaeologist John Wymer. By this point his relationship with the South African Museum had deteriorated. Singer had clashed with Brett Hendy, the excavator at Langebaanweg. When the museum refused to give any formal recognition to Singer's coloured excavator, Saunders, he decided that he had had enough. He gave up the Klasies excavations and moved to the Middle Stone Age sites of Hoxne and Clacton in England, relocating both Wymer and Saunders at his own expense. It had not helped that his personal relationship with the director of the South African Museum, Tom Barry, had also worsened. Singer had no compunction about borrowing specimens from South Africa to study in Chicago. He had done this with much of the Department of Anatomy's historical archive and several cadaver crania, but he also had taken the cranium from Saldanha (Elandsfontein) and the Fish Hoek skeletons from the South African Museum. Barry had written progressively more acrimonious letters to Singer asking for their return, but Singer kept delaying, saying that he had not finished his research on them. In the end Barry had to go to Chicago and literally take back the specimens by hand.[33]

Singer could be a difficult man. He not only clashed with Dart, Hendy and Barry, but also with his colleague John Day in zoology at UCT. Although he spent a lot of his time advising postgraduate students, few of these worked with him in the Department of Anatomy as most were either from archaeology or zoology. Alun Hughes thought he was a strange chap: 'Singer is manic … very good, very clever, but slashes people to bits when he wants to and I don't think he gets on too well with people like Phillip Tobias and didn't get on well with Dart.'[34] Although Singer published frequently in human biology and palaeo-anthropology and archaeology and he developed a strong group of students in Chicago, he had little influence over the development of South African anthropologists after 1962. Some of his research data remain unpublished and on his death in 2006 his papers were split between the University of Colorado in Denver and the Arizona State University in Phoenix.[35]

THE 'NEW PHYSICAL ANTHROPOLOGY'

Comparing the early careers of Tobias and Singer brings to light several experiences that they had in common. Both were exposed to the study of Mendelian genetics, although neither was taught by geneticists. Singer approached the field through his interest in developmental embryology; Tobias had to teach the basic tenets of the subject to his supervisor Joe Gillman. Both took part in expeditions that exposed them to a holistic view of people in their ecological and social context. Tobias's mentor, Dart, had also done fieldwork in a sense, but he had not understood the concept of studying people as populations, rather than types. Tobias and Singer both had early career opportunities to travel to research institutions overseas. Singer was based in a department of embryology, but he did visit other medical schools and centres of anthropology in the United States and England. Tobias had a more formal experience and was more strongly involved with physical anthropological research, but both he and Singer were attending international conferences on a regular basis by the late 1950s. Lastly, both Singer and Tobias were aware of and engaged with political issues that impacted on teaching and research in South Africa.

Physical anthropology in the scientific community after the Second World War was a field in flux. There was a sense of shock after the exposure of the atrocities committed by the Nazi regime in its homeland and in its conquered territories. Scientists in Germany had not only collaborated with their political masters, but had also been central to the implementation of their racial hygiene policies. Even worse was the realisation that the eugenic policies proposed and implemented in the Western world were only different in degree from their German cousins. There had been a group who opposed the pseudo-genetic models of racial hierarchy present in pre-war anthropology. Julian Huxley, Alfred Cort Haddon and Alexander Carr-Saunders published their anti-racist *We Europeans* in 1935 and Lancelot Hogben translated Gunnar Dahlberg's *Race, Reason and Rubbish* from the original Swedish in 1942. The same year also saw the publication of Ashley Montagu's book *Man's Most Dangerous Myth.*[36]

Huxley's *Evolution: The Modern Synthesis* came out in 1942. In this publication Huxley explored the relatively new field of population genetics in relation to evolution. This was a critical volume because it united Mendelian and biometrical analysis through population genetics, although it did not specifically discuss the issue of human races. The marriage between Mendelian genetics and biometry was never a sound one. Although the biometricians could effectively display real differences between sample populations, they could not provide a theoretical model of inheritance in their statistical analysis of variation. The theory rested in the hands of the Mendelian geneticists who found typology a much more useful tool for their proposed social engineering. The publication of Ronald Fisher's multiple gene hypothesis in 1930 was a way in which the biometricians and the Mendelians could be reconciled. By removing simplistic approaches to genetic inheritance, Fisher opened the field of population genetics and showed how particulate inheritance could produce a continuous range of variation (Magner 1979).

The scientists realised that at the heart was the problem of the definition of race itself. A group of anthropologists convened a conference in Paris under the Unesco banner and issued a *Statement on Race* in 1950. The committee was primarily made up of social anthropologists and only Juan Comas and Montagu had a physical anthropology background. Among the points made in the Statement were that there was no such thing as pure races and that all human populations were of one species, which could interbreed with no deleterious effects. No correlation was found between race and behaviour and race was identified as a sociological and not a biological construct (Spencer 1997c, 1052). The Statement was heavily criticised because of the lack of physical anthropology opinion, and in response a new committee was convened in Paris the following year and the Statement was reconsidered. The revised Statement, published in 1951, did not differ substantially, but it was expanded and much of the ambiguity was removed. Key to the 1951 Statement was the recognition that racial categories were a mechanism for understanding the historical differentiation of populations, but it is the population, not the race, which is the evolutionary unit (Hiernaux 1969, 11). Debate continued after the publication of the second Statement, with many physical anthropologists arguing that it was implying that racial variation did not exist. Typology remained the technique of choice in most descriptions of human variation. Why was this so? Few physical anthropologists had the necessary training to understand the complex mathematics involved in biometrical anthropology. Nancy Stepan (1982, 138) has also proposed that because race history was still the major focus of anthropological research, scientists did not want to leave the easily understandable shores of typology. Racial classification was for most anthropologists the whole point of their measurements and biometricians were not perceived to be doing racial analysis.

The break from the old typological analyses accelerated after the Unesco statements on race. The critique of race found fertile ground in the American schools, which taught the broad concept of anthropology. Linguistic, cultural, archaeological and physical fields intersected in American teaching, unlike the European and South African programmes in which these subjects were separate (Hultkrantz 1980, 89). Sherry Washburn at the University of Chicago was the first to launch a change in approach in the study of human variation, which he called the 'new physical anthropology' (Washburn 1951). Simply put, he argued that the old physical anthropology was constructed around the study of races and it was essentially descriptive and static. What was needed was a study of populations, not races, in which the process of dynamic change was the focus of the study and description was no longer the objective (Spencer 1997b, 1104). Physical anthropology had to adopt the methods and ethos of modern evolutionary theory, especially population genetics. It needed to study how gene frequencies change through migration, drift and selection (Washburn 1951, 299). It was, he argued, pointless to study races because the reality of human evolution was that populations were always changing. Washburn's view was spread by his students and by the early 1980s less than half of the departments of anthropology in the United States considered race as a topic for study or research (Littlefield, Lieberman and Reynolds 1982, 646).

In Great Britain, the lance carrier of the 'new physical anthropology' was Joseph Weiner. Weiner had been one of Dart's science students in Johannesburg, graduating with a Bachelor of Science degree in 1934, six years after Wells. Unlike Wells, Weiner continued in physiology, rather than anatomy, gaining experience studying heat stroke and heat exhaustion in black miners at the Witwatersrand mines (Little and Collins 2012, 115). Relocating to England in 1937, Weiner completed a medical degree and a doctorate from the University of London. He took the post of reader in physical anthropology in the Department of Anatomy at Oxford University at the end of 1945. Under Wilfrid Le Gros Clark, the department at Oxford had moved away from traditional craniological studies and had transferred its collection of archaeological and ethnographic human skeletons to the British Museum of Natural History. Le Gros Clark wanted something different and hired Weiner specifically to generate a new approach to physical anthropology (Roberts 1997, 1108). This is exactly what he did, beginning a series of studies of communities as dynamic entities adapting to stresses and demands of the environment. He saw physical anthropology as a living discipline that was relevant to modern scientific problems. As to race, he saw the complexity and variability of human populations as meaning that races were abstractions, which could only be described statistically (Weiner 1952, 408). From his base in Oxford, Weiner collaborated with both Tobias and Singer in expeditions to study the Khoesan people in Bechuanaland and South West Africa. His interest came out of his earlier physiological studies of heat stress and he had become interested in the evolutionary process of physiology and climatic adaptation. Within a few years this would become recognised as a new field of human ecology. In 1962 Weiner was named world convener of the Human Adaptability Section of the IBP (Meier 2018). He would draw Tobias into this programme. Among the IBP projects was a comparison of similar populations in different environments, which Tobias would organise for farm (herdsmen living with their families on European farms near Ghanzi) versus wild (groups living as foragers) San in the Kalahari, and for town versus country among the Venda and the Pedi in the northern Transvaal.

The influence of the 'new physical anthropology' can be seen in the writings of both Singer and Tobias as they switched from typology to what would become known as human biology. In his Nusas lecture of 1955, the closest Singer ever got to publishing a political paper, he described all the elements of the new physical anthropology without using its name. He talked about the integration of population genetics into the study of anthropology and noted that pure races did not exist except as theoretical racial types. He further remarked that although race classification was a useful technique in cataloguing, the type tended to 'dominate the maker' and the concept was incompatible with Mendelian genetics (Singer 1955, 62). Three years later he published his destruction of the Boskop type with the often-quoted comment: 'It is now obvious that what was justifiable speculation (because of paucity of data) in 1923, and was apparent as speculation in 1947, is inexcusable to maintain in 1958' (Singer 1958, 177). As his studies of living populations increased, Singer began to think of the

Khoesan populations as part of the broad spectrum of African peoples, not as individual types but as Mendelian populations interbreeding with one another. This perspective developed even more once he had begun to work with Weiner and they talked of human populations as never being static in biological make-up, 'and even over quiet short periods they evince those changes termed evolution' (Singer and Weiner 1963, 168).

Unlike Singer, who was little influenced by the typology of Drennan, Tobias struggled to free himself from the shadow of Dart. Writing in 1953, at the height of his political activism against the newly imposed apartheid legislation, Tobias outlined his understanding of the concept of race. He refers to the Unesco *Statement on Race* and echoes their discussion of the genetic nature of populations and the arbitrary nature of racial classification, but continues to refer to physical types as a subdivision of racial variation. Most importantly, he stresses that variation makes the classification of single individuals difficult and 'no physical anthropologist could set himself up as a scientific arbitrator over these boundaries in a multi-racial community like that of South Africa' (Tobias 1953, 123), a clear reference to the implementation of the policy of apartheid at the time. He continued to struggle with the confusing ancestral types in South African archaeology (Tobias 1955, 10; 1959, 139), but successfully made the transition with the publication of his pamphlet *The Meaning of Race* in 1961. This pamphlet, and its revised and expanded second edition in 1972, covers much of the same ground that Tobias attempted in his 1953 paper, but the argument is clearer, and he has jettisoned the last of the typology.

Perhaps the best way to see the impact of this transition is to compare similar chapters by Dart and Tobias for different editions of the same book. In 1937 Dart contributed a chapter on racial origins for Isaac Schapera's *The Bantu-Speaking Tribes of South Africa*. It is very much Dart in his typological prime, expounding on skull shapes, migrations and racial attributes. In 1974, W.D. (David) Hammond-Tooke launched a second edition, with the slightly different title *The Bantu-Speaking Peoples of Southern Africa*, and invited Tobias to contribute the introductory biological chapter. Although Tobias covered the exact same ground as Dart, talking about the various tribal groupings and historical subdivisions, there are few other similarities. Instead of characters revealing the essences of type, Tobias's chapter looks at the whole range of people in their biological context. Descriptive morphological data remain, but they are ranges of variation rather than averages. These data are joined by blood biochemistry and nutrition information, along with discussions of gene frequencies and adaptations to local ecology.

Tobias articulated the concept of this departure particularly well in his chapter for Hammond-Tooke's book: 'The new approach stresses rather the environmental or selective pressures which have led populations to diverge in genetic compositions. It is concerned with the adaptive benefits, the usefulness, the survival value, the selective advantage or disadvantage, of a particular gene or set of genes. Thus, it sees man in his environment, physical, biotic, social' (Tobias 1974, 4). Although Tobias, like Singer, never mentions it by name, this is the 'new physical anthropology' and it represents a significant break with the past.

THE WITS DEPARTMENT OF ANATOMY UNDER PHILLIP TOBIAS

Tobias inherited an academic department that Dart had taken 35 years to build. Although he had been associated with it for more than ten years himself, the thought of taking over was daunting. By his own admission, he was reluctant to take up the post. Dart had insisted that he apply for Drennan's chair at UCT in 1955, even when he protested that he was not ready and he was proven right when the chair was given to Wells instead. He still was not confident in 1958 and had not put in his application for the Johannesburg job as the closing date approached. When Dart found out, he made Tobias 'sit down and write out [his] application without delay' and it was handed in on the closing date (Tobias 1991, 18). In truth, the selection committee had a tough decision to make, as there were three highly qualified applicants shortlisted: Weiner, Singer and Tobias. In an unprecedented move, the committee asked Dart for his opinion. Dart asked Hughes,

> 'Who would you like to see as Head?' Without giving it a thought, you know, after he'd cited a few names, I said, 'Singer'. Of course, he made an impassioned plea for Tobias and Tobias got the job. You see he didn't want to lose [his legacy], he was afraid. You see, Singer would have destroyed a lot of his work, and Weiner would have thrown it all out.[37]

Dart clearly saw Tobias as being in his mould and, although there would be changes, Dart felt that Tobias would provide continuity and keep to traditional anthropological fields. While the selection committee was busy deliberating in October 1958, the University of Michigan at Ann Arbor offered Tobias a double chair in anthropology. Tobias chose Wits when the selection committee announced its decision to him, partly for family reasons and because he wanted to continue his African research in Africa, but also because he saw Wits as a shining light fighting against apartheid (Tobias 2000, 37).

Tobias's management style was very different from Dart's. After his return from medical leave during the war years, Dart had delegated nearly all his departmental duties to his staff and spent most of his time writing in his office. Hughes noted that Dart was always available and if you needed to speak to him, you just knocked on the door and went in. That stopped when Tobias took over, Hughes remembered: 'There was a secretary. There was a barrier between the staff and the Head of the Department and delegation came to an end.'[38] Staff meetings, which had been non-existent under Dart, were long and formal under Tobias. His relationships with staff also tended to be more formal. Jenkins found Tobias autocratic and authoritarian when he worked in the department in the 1960s (Kuljian 2016, 148). De Villiers noted how Tobias expected to be addressed formally in an academic setting even before he took over the headship. As a technician she travelled with him to do fieldwork at Makapansgat. The entire crew stayed at the research accommodation on site and everyone was very friendly. He invited her to call him 'Phillip', but when the trip was over and Tobias was dropping her off in Johannesburg, he said, '"Hertha, remember that tomorrow I am Dr Tobias." Since that day, and it was a long time ago, I have never called him Phillip, it has always been Dr Tobias or Professor Tobias.'[39]

Hughes felt that not only was Tobias more unapproachable than Dart, but he was also not as social and that he neglected to introduce staff to visitors to the department.[40] During my own time in the department between 1975 and 1979, I found the situation very different (Morris 2010a, chapter 9). Tobias would go out of his way to make sure that the doctoral students met all the visiting scientists and often arranged for social occasions with them. One of Tobias's favourite haunts was the Gramadoelas restaurant in Hillbrow. Gramadoelas, meaning a place that is far away and remote in Afrikaans, specialised in South African foods. The restaurant was just down the street from Tobias's apartment, and not only did Tobias know the owners well, but they always made a big fuss of him. The party, often consisting of the foreign guests and several of the postgraduate students, would be there as Tobias's guests. Another Tobias specialty was the departmental party. Whenever there was an occasion worth noting, Tobias arranged to have a gathering either in the departmental tearoom or in the museum. Chief among these occasions was Dart's birthday on 4 February. These did not always work out the way he had planned. Dart liked to tell the stories of the old days, but these became a bit confused as his memory faded, and sometimes he needed a bit of prompting from Tobias to remember a detail or two (Morris 2010a, 84).

The luncheon club was another social innovation by Tobias that was an attempt to unite the whole department. He had attended a weekly staff luncheon at Yale University and had been very impressed with the idea.[41] He initiated a similar gathering when he was in charge in Johannesburg. All staff, including the African technical staff, were required to attend, and it provided an opportunity to socialise in a context that would have been difficult outside of the university. Tobias tried to get as wide a range of speakers as possible, not just in science. The talk in one memorable luncheon club in 1975 was given by Percy Qoboza, who was then editor of the newspaper *The World* in Soweto. The topic was political and unapologetically unsupportive of the South African government.

Tobias's sociability extended to the undergraduate students as well, but he did try to keep a social distance between himself and them. He would take the whole of the second-year medical class out to the twin sites of Sterkfontein and Swartkrans each year as a field trip linked to the anatomy course. These annual outings were an opportunity for Tobias to become one of the gang and he would often allow students to engage in good-natured high jinks. It was all in good fun, but it did not last beyond the one day. The May 1975 excursion was particularly memorable because the high jinks were higher than usual. The group was especially boisterous, and a few students did a naked streak across the roof of the Sterkfontein tearoom. Word of the rowdy behaviour made its way to Revil Mason, of the Department of Archaeology, who at that stage was not getting on with Tobias and he chose to make a formal complaint to the vice chancellor about the matter.[42] The dean of the Faculty of Medicine demanded that Tobias clarify the events of the day and provide disciplinary action or apologies as required.[43] Tobias was furious and gave a serious dressing down to one of his doctoral students who had reported the events to Mason (Morris 2010a, 76–77).

Presumably Tobias apologised to the dean, but he chose not to penalise the undergraduate students.

The postgraduate science students had a special place in Tobias's department. Geoff Sperber estimated that Tobias had trained 32 doctoral students (Sperber 1990, ix) and although there is no formal list, a search through the records suggests his career total is about 35 graduating students, split between the faculties of science and health sciences, including some who were supervised by him well into his retirement. The largest single group worked on projects involving the fossil hominids, but more students in total studied modern human variation, growth and clinical anatomy, reflecting Tobias's self-identity as a human biologist (Thomas and De Bruyn 1985, 5). Under the South African/British system of supervision, the relationship between supervisor and student tends to be very close and personal, which means that not all of Tobias's supervisions were without conflict. He tended to take relatively few students at a time, but in the late 1970s he accepted a cluster of foreign postgraduate students. He tried to have a weekly three-hour session with the student group, but found this too time intensive and instead requested that students come to see him if and when they had problems.[44] This is consistent with what I experienced during my time in the department when communication with Tobias was often done with notes, rather than face to face (Morris 2010a, 83). What was notable was that Tobias was little engaged with the construction and layout of the projects as the student began his or her studies, but was much more involved once chapters were being written (and frequently rewritten). Completing a doctoral thesis under Tobias was almost invariably a slow process.

De Villiers was Tobias's second doctoral supervision. He was initially asked to act as internal examiner for her Master's degree, but took over the role as her supervisor when Dart retired at the end of 1958.[45] De Villiers had intended to work in neuroanatomy for her doctoral thesis, but Dart had persuaded her to continue with the cranial variation project she had begun with her Master's. She continued in much the same line in her doctorate, but used more elaborate statistics, including the still new Penrose size and shape calculations, which she did by hand over a two-year period, using a Facet calculating machine.[46] She drew heavily from the biometric school of analysis, but struggled with Tobias's supervision because he insisted on her engaging with typological assessments. One point of contention was over Boskop. Tobias wanted it included as a concept, but she disagreed. He forced her to include a section on Boskop in the thesis, but it was removed for the published version (De Villiers 1968). Tobias, as internal examiner, took more than a year to examine the final submission although his two external examiners, Jack Trevor and T. Dale Stewart, were much more attuned to statistical analyses and responded relatively quickly.[47] How innovative De Villiers's thesis was for South African anthropology is evident in comparison to research being done just ten years earlier. Wells had begun work in the late 1940s on his own book and had still been working on a draft when he came to Cape Town in 1955. Whether he stopped work on his project because he discovered De Villiers was engaged in a similar work is unknown, but the Wells manuscript shows that he was using the old typological assessment techniques and the study was almost free of statistics, other than frequency tables.

Hertha de Villiers (1924–2017) was born in Johannesburg, but moved to Cape Town with her family when she was nine years old. Upon completing high school, she enrolled in a Bachelor of Arts language programme in 1942 at the University of Stellenbosch on the suggestion of her father, who felt this would improve her Afrikaans language skills. After only a year she switched to UCT to do a basic science degree in preparation for medicine, graduating in 1945. Her majors were zoology and physiology, but she included Drennan's medical anatomy course in her studies. Again, she deferred to her father's wishes, agreeing that it would be better if she did the Bachelor of Science degree before she went on to do the MBChB.[48] She was employed as a researcher in the Bilharzia Research Unit of the South African Institute for Medical Research in 1949, but two years later joined the neighbouring Department of Anatomy at Wits as a technical assistant.

The job in anatomy brought her into contact with Dart and it was Dart who would have a major influence on her academic future. De Villiers's plan was to enter the third year of the medicine programme using her Bachelor of Science degree as credit for the first two years. When Dart heard this, he persuaded her to pursue science instead:

> That was when he got hold of me and he said, 'What do you want to do medicine for? I'll help you, you know I'll assist you toward a higher degree, so stay in Anatomy. There is so much to do in this field. There is this wonderful collection of skulls and skeletons that I've collected over 30 years or so.'[49]

The technical post that De Villiers accepted required her to teach third-year Bachelor of Science students in techniques of casting and modelling, but she was also responsible for the accession and cataloguing of anatomical, anthropological, archaeological and zoological material into the anatomy museum. Under Dart's supervision, she completed her Master's in 1957 and was promoted to the academic stream in the department as a lecturer. Dart continued to pressure her, this time to complete a doctorate, which she completed in 1963, but De Villiers was disappointed that further promotion was slower than she had hoped. She applied for advancement to senior lecturer after the acceptance of her doctorate, but it took four years for this to come about. De Villiers's academic role in the department was similar to that of Galloway and Wells during the Dart years. Along with her second-year medical student teaching duties, she was involved in most things anthropological. The task of teaching research methods to the science students was hers and she supervised nearly all of the field projects that the students took part in, although this became more difficult as her working partnership with Tobias began to fail. She was promoted to associate professor in 1972, but chose to leave the department two years later to take the headship of the Department of General Anatomy at the Wits dental school. This did not come with a promotion to full professor, but it did mean she could escape the deteriorating academic relationship with Tobias.

The range of research topics being carried out in the Department of Anatomy at Wits reflected Tobias's broad vision of human biology. He had become deeply involved in palaeoanthropology from 1959 when Mary and Louis Leakey had invited him to study their newly discovered hominin cranium from Olduvai, OH5 (*Australopithecus boisei*). Tobias would continue to work on the Olduvai hominins with his seminal study of the *Homo habilis* remains, but from 1966 he reopened the Sterkfontein type site nearer to home. Work continued intermittently at the Makapansgat Limeworks site and his senior lecturer Jeff McKee focused on the Taung site near Kimberley, which had been neglected since the discovery of the original Taung specimen in 1924. Tobias's renewed focus on the South African australopithecine sites generated a new group of researchers who would carry on his palaeoanthropology programme right up to the present day.

Palaeoanthropology certainly brought the lion's share of publicity to the department, but Tobias continued in Dart's tradition by spreading his interests into many other areas (Štrkalj 2016, 1). Tobias continued the primatology research on the chacma baboon colony that Dart had started (Štrkalj and Tobias 2008, 275), which generated significant physiological and clinical research. Tobias, in collaboration with Dart, extended this to another of the living primates. Dart had always had a special interest in the gorilla after his 1930 expedition to the Congo. When Walter Baumgartel of the Traveller's Rest lodge in Uganda contacted him in 1956 to help with funding gorilla research, Dart jumped at the chance and arranged for Tobias to visit Uganda in June 1957, followed by his own trip to the Congo and Uganda in November.[50] Dart and Tobias approached the Wits research committee and were able to convince them to fund a Witwatersrand Uganda Gorilla Research Unit to support research being launched from Baumgartel's rest camp at Kisoro (Tobias 1961b, 297). In 1958, Niels Bolwig, one of Dart's senior lecturers, initiated field research in Uganda, setting up camp on the saddle between the Virunga volcanoes near Kisoro. Over a period of two months he tracked gorillas and although he was not able to get close enough to make systematic observations, he did gather data on their nightly nest constructions. His field notes helped George Schaller set up the more successful mountain gorilla project the following year (Petersen 2004). Bolwig joined Galloway, the dean of medicine at Makarere University in Kampala from 1959, and the two of them kept Dart and Tobias informed about developments in gorilla research. Funding from the Wits university research committee was stopped in 1960 because of political pressure against South Africa.[51] The Uganda wildlife department refused to allow gorilla skeletons to be sent to South Africa for study and Bolwig left shortly after to take up a position in Nigeria (Bolwig 1988, 53). When Galloway retired in 1962, his replacement, David Allbrook, changed the research direction of the Makerere anatomy department and ended gorilla studies there.[52]

Back in South Africa, Tobias was developing a substantial interest in the living populations of southern Africa. The KRC was launched just after his return from the Rockefeller Fellowship in the United States. The Panhard-Capricorn Expedition, of which Tobias had been a member, was among the first, but the Marshall family from the United States were doing their own ethnographic studies at the same time, and A.J. Clement of Wits had been in the southern Kalahari gathering dental data on populations of 'Bushmen' in August 1951. The proposal to convene the KRC was suggested because there was little or no co-ordination between these studies. With Tobias as chair-person, the committee mounted 21 expeditions between 1958 and 1967 (Tobias 1975, 74). Most of these expeditions included growth and clinical studies and medical treatment. The object was to gather as much information as possible about the people visited in order to understand their ecology, local adaptation and range in variation of physiological and anatomical features.

De Villiers was a member of one of the earliest expeditions to the Kalahari. She was co-opted in the absence of Tobias to gather meas-urement data on the 1958 expedition led by Weiner. There were objections when she was nominated by Tobias because she would be the only female. The solution was to bring a second woman along who would work with De Villiers. In the end, it was only the two female researchers, along with the psychologist Helmuth Reuning, who completed the trip right up to the Okavango Delta. De Villiers was witness to the methodology of gaining the co-operation of the San groups contacted. The San were notified by the district commissioner to visit the research camp. Mealie meal and game meat were provided, and the people stayed as long as they were fed. Culture clashes were evident when the physiologists began to collect electrocardiograph data and urine samples. The connection of the electrocardiograph machine via cable to the battery of the truck was a frightening con-cept, especially because the skin of the subject had to be cleaned with alcohol and methylated spirits beforehand. The people were also distressed that the urine samples from different individuals were discarded together in a single refuse hole when the tests were com-plete.[53] Body fluids such as urine were intensely personal to the San and would never have been mixed in their society.

The Kalahari projects were just one aspect of Tobias's studies of living populations. He collaborated with Weiner on the IBP studies on human variability. Not only did he structure some of the Kalahari San data into 'farm' versus 'wild' (as defined earlier in this chapter), but he also arranged for De Villiers to work on similar paired groups of urban and rural Sepedi- and Tshivenda-speaking people. Tobias had arranged for growth studies of the San children and later, in the 1960s and 1970s, he extended these to include groups of Bantu-speaking children and children in the Chinese school in Johannesburg.

The intensity and invasive nature of this data collection did not raise any issues at the time, but in hindsight several researchers have expressed concerns about the disjunction between Tobias's political vision of humanity and his scientific detachment. Catherine Burns (1996, 10) was extremely concerned that observations of external genitalia were still being done as part of physical descriptions in the late 1960s by Tobias and his co-workers. She is certainly correct in noting that the earlier researchers took a most unhealthy interest in features sexual and she is highly critical of Lawrence Wells and Margaret Orford in their studies of 'Bantu' women. She questions why these features were recorded in a study that was intended to complement obstetrical knowledge when such features were just ana-tomical flags for racial typology. Several studies put sexual features as primary racial characters and devoted whole papers to their clas-sification and illustration, especially of the female variations (Drury and Drennan 1926; De Almeida 1956). When Dart published his description of the ‡Khomani San (Dart 1937a), he had included photographs of men displaying their semi-erect penises (ithyphally), and women displaying their enlarged labia minora (macronymphae). The pictures were printed as separate pages, which could be inserted into copies sent to a select few and removed from the general print run. Tobias saw it as of interest enough to ask Carol Orkin and De Villiers to gather data about the labia minora in their study of the Griqua (Orkin 1970, 42). This was allied to the human adaptability section of the IBP set up by Weiner, and not a continuation of Dart's programme of field studies of living peoples as proposed by Burns

(1996, 9). Tobias included records of both male ithyphally and female macronymphae in his list of soft-tissue physical characteristics of southern African peoples, along with hair form, eye folds and earlobe shape (Tobias 1974, 22). This suggests a possible reason for why he continued to look at features of external genitalia. Tobias was not viewing these genital variations as peculiar. He was simply enumerating physical features, which have different frequencies in different southern African populations. For Tobias, they were just two variable characters of many physical features, not racial markers, and of much less value than the genotypic variability being mapped by the serogeneticists. Tobias did see the political connections between biology and society in relation to the big picture of racial inequality in South Africa. He proposed the concept of positive, negative and absent secular trends to account for the effects of nutrition on stature and how oppressed and marginalised groups were not demonstrating the general increase in body height that was evident in other world populations (Wood 1989, 221).

The other aspect of human variation that concerned Tobias was skeletal anatomy. The Dart collection of human skeletons from the dissection hall was an important resource. Tobias's students who were working on ancient bones from the fossil hominin sites were requested to gather comparative data on modern humans from the skeleton collection. His more medically oriented students studied the range of variation of specific skeletal elements of clinical importance. Despite Tobias's special interest and field experience in archaeology, I was the only one to choose a doctoral topic that involved archaeological skeletons (Morris 1984), although Pamela De Beer Kaufman (1975) made use of the Khoesan skeletons from archaeological contexts in Kimberley and Johannesburg. De Villiers's doctoral project dealt with the cranial variation of modern populations (1963), but she regularly worked on excavated Holocene archaeological material after finishing her doctorate.

Tobias did initiate one project that required the archaeological excavation of graves. In 1961 he was approached by Basil Humphreys to help identify the grave of the Griqua leader Cornelis Kok II. Humphreys owned a farm near Campbell (in what is now the Northern Cape) and had become very interested in the history of the Griquas resident in the area. He had become a friend of the leader of the Campbell Griqua community, Kaptein Adam Kok IV, and it was with his blessing that Humphreys approached Tobias with the idea of excavating some of the graves in the local Griqua cemetery, with the object of confirming the grave of Cornelis Kok II.[54] Over four field seasons (1961, 1963, 1967 and 1971), Tobias engaged groups of his science students to help excavate the graves and ultimately 35 skeletons were accessioned to the Dart collection in Johannesburg (Morris 1992b, 59). Tobias did not attempt a study of the skeletons after excavation and a description of these people had to wait for my own work some twenty years later.

Christa Kuljian (2016, 183) raises the issue that Kaptein Adam Kok IV did not fully understand that Tobias would keep the skeletons for as long as he did, although ultimately in 2007 the skeletons were ceremoniously reburied at the grave site. The long delay left Tobias open to the accusation that he was keeping the skeletons in secrecy (Kuljian 2016, 219) although Tobias never hid the fact that the skeletons were in the department. Perhaps the reason for the long delay was that Tobias was less interested in the skeletons than he was in the community that they represented. He renewed his focus on the living community in 1971 with a study of body form, fingerprints, taste-testing, colour vision and childhood growth. De Villiers did most of the measurements, but she was joined in December 1971 by George Nurse to gather genetic data from blood and urine samples. This resulted in a clash between De Villiers and Tobias, who accused De Villiers of stealing his project. She refused to participate in any further field projects and soon after she moved to the Department of General Anatomy at the dental school. When De Villiers retired from

Wits, she 'destroyed all of the correspondence that I had had with Professor Tobias because much of it was very acrimonious and really quite nasty'.[55] Nurse and Jenkins continued with the genetic research, producing several papers that made the Griqua community one of the better-understood genetic populations in South Africa (Nurse and Jenkins 1975; Nurse 1976).

De Villiers's conflicts with Tobias did not represent most of his relationships with his colleagues and students, but his strong will did mean that such difficulties occasionally occurred. De Villiers was particularly angry because Tobias did not support her promotion beyond

Figure 6.3: Excavating Griqua graves in April 1961 at the old Campbell cemetery in the Northern Cape. The photograph shows Gerhard Fock (standing in grave) and behind him, from left to right: Phillip Tobias, T.R. Badenhorst, P.J. Olmesdahl (students) and Tad Kaufman (technician). (Source: Gerhard Fock, courtesy of Dora Fock)

the level of senior lecturer and over the years several highly competent and skilled members of his department left to find promotion
elsewhere when Tobias blocked advancement to a professorial level.
It is highly unlikely that this failure to promote anatomically and
anthropologically oriented staff was the result of a failure to match
his own high standard and it is more likely that Tobias did not enjoy
competition at a senior level.

Except for his brief flirtation with the Liberal Party, Tobias did not
join any political organisations after he stepped down from the Nusas
presidency, but this did not mean that he stepped down from his political viewpoints. He was still a staunch enemy of apartheid, especially
when it came to the application of apartheid in the university, and he
was a frequent speaker at public gatherings where these political issues
were discussed. He addressed an assembly of 2 000 students in April
1969 on the tenth anniversary of the proclamation of the Extension
of University Education Bill, which racially segregated the universities
(Shear 1996, 33). Tobias also spoke before an angry student assembly
in October 1977, protesting the death of Steve Biko and the banning
of eighteen organisations mostly supporting the Black Consciousness
movement (Welsh 2009, 104). When the government tried to shift the
administration of university apartheid from the ministers to the universities themselves in the Quota Bill of 1982, Tobias was again prominent
among those protesting publicly: 'Tobias told the Assembly that the
Senate found the Minister's conditions totally unacceptable and declared
that its members would decline to comply with them' (Shear 1996, 185).

Tobias was most vociferous when it came to medically qualified
doctors who co-operated with the apartheid state. He, along with
many others, was saddened by the death of Biko in police custody in
September 1977. Three medical practitioners had allowed the police to
transport Biko to Pretoria in the back of a police van despite the life-
threatening injuries he had sustained during interrogation. Tobias was
furious when the South African Medical and Dental Council (SAMDC)

Figure 6.4: Raymond Dart's 84th birthday celebration at the Department of Anatomy, University of the Witwatersrand, February 1977. From left: Fred Grine, Raymond Dart, Ivan Suzman, Ron Clarke, Phillip Tobias and Alan Morris. Dart had retired from Wits in 1959, but Tobias celebrated his birthday every year with a small gathering in his honour. (Source: Photograph by Peter Faugust)

decided at a disciplinary inquiry in 1980 that no prima facie evidence existed of the improper conduct of the three doctors. The board of the Wits Faculty of Medicine refused to accept the finding and disassociated itself from the decision, but Tobias was to take it further. Tobias was dean in November 1984, and he, Jenkins and Francis Ames of UCT took the matter to court and forced the SAMDC to hold a new inquiry. Benjamin Tucker and Ivor Lang were found guilty, but given very light sentences. Tobias and the opposition member of Parliament, Helen Suzman, objected publicly and finally the SAMDC struck Tucker's name from the medical register (Shear 1996, 291).

Tobias and a number of South African archaeologists and palaeoanthropologists were excluded from the International Union of Prehistoric and Protohistoric Sciences' (IUPPS) eleventh conference, scheduled for Southampton in September 1986. The decision was made by the local organising committee as part of an academic boycott in support of the anti-apartheid movement. South Africans had been excluded from conferences before this, but it had been the decision of the host nation and its visa policy, not the conference organisers themselves. The parent body of the IUPPS had no political policy in place to enact such a boycott and Tobias was vigorously opposed to its implementation. By opposing the boycott, Tobias was not supporting apartheid. Instead, he was supporting an ideal of 'the long-recognised and respected principle of the universality of science and scholarship' (Tobias 1985b, 670). Tobias agreed with the organising committee that the policy of apartheid was abhorrent, but he felt that the boycott, no matter how laudable, hampered the principle of the free circulation of scientists. This, in Tobias's opinion, was a precious freedom that was for the greater good of humanity and went beyond politics. He warned that such activities would damage the already fragile university structures in South Africa by increasing the emigration rate of scientists and damaging the speed of recovery once apartheid finally expired (Tobias 1985b, 667). He saw a boycott as entirely negative (Wood 1989, 233) and felt that the meeting of scientists would be an opportunity for all of them to make resolutions against apartheid without disturbing the flow of science.

The IUPPS committee met in January 1986. They removed the Southampton organising committee and moved the eleventh conference to the city of Mainz in Germany under a new committee and with South African and Namibian scientists included.[56] In response, the Southampton committee withdrew from the IUPPS and went ahead with their own September conference, now called the World Archaeology Congress. Peter Ucko, who was the secretary for the Southampton organising committee, rationalised the position as being necessitated by the unique situation in South Africa, where the academic boycott was necessary to build pressure on the South African government to remove apartheid. He sympathised with the scientists in the IUPPS, but believed that the collateral damage to science was necessary for the higher good. He could not understand how people like Tobias could continue to attend international conferences when this was 'directly contrary to the ANC's position' (Ucko 1987, 116). Interesting here is the juxtaposition of a political party (the African National Congress, ANC) with the more general opposition to apartheid. When the Union of Democratic University Staff Associations (Udusa) was formed in 1987, one of the proposals was that Udusa would set up a committee that would vet abstracts from scientists who wished to participate in overseas conferences. Thankfully, this misconceived idea was not enacted. It would have been the ultimate destruction of academic freedom, with political parties and unions making judgements on the value and direction of research. Although the academic boycott continued, it was up to individual conferences to make their own decisions as to whom to accept and whom to reject.

The present prevailing belief is that the academic boycott did indeed speed up the demise of apartheid, but the damage that Tobias feared has never really been assessed. The IUPPS has continued to meet every five years, while the World Archaeology Congress generally meets every four years. I attended World Archaeology Congress 4 (Cape Town), 5 (Washington) and 6 (Dublin) and noted that all three of these meetings kept their political focus. At the Washington meeting I was impressed with how the South African delegates were trying to bring archaeology to the service of descendant communities, while the Australians and Americans remained polarised in their finger-pointing about blame from the past (Morris 2004, 4). Certainly, the wish of the organisers to change the political and regional focus of archaeology has been successful, but the price paid is that relatively little advancement of knowledge has occurred in comparison with that reported at the Society of Africanist Archaeologists and the Association of Southern African Archaeologists at their biennial meetings. Perhaps one could say that there is a dual legacy for archaeology in southern Africa: one from the World Archaeology Congress, which has sensitised science to its social context, and a second from Tobias, which has built pride in the quality and value of the research.

The start of this chapter noted how Tobias was concerned about how his legacy would be seen in the future. My linking of Singer and Tobias was an irritant because Tobias thought his role was much more important than Singer's. In fact, Tobias was right. Although Tobias and Singer both crossed the bridge from typology to the 'new physical anthropology', it was Tobias who publicly spoke out against the indignities of apartheid and how it impacted on South Africa's education and medical practice. He stayed in South Africa and was watched closely by the Bureau of State Security. He challenged the apartheid masters in court, but more importantly he sensitised his students to politics. Unlike Dart, who pretended that science did not involve politics, Tobias guided his students by his actions. Even when he did not support the politically correct version, such as in the Southampton case, he was explicit in his reasons, disagreeing with but not dodging the issues. Tobias spoke about his academic grandchildren, each of whom went on to mentor their own students (Wood 1989, 222). The connection with people – students and the interested public alike – is his true legacy. It is not just the thousands of medical students he taught, or the 35 doctorates that he supervised; it is also the thousands of South Africans who know his name, his scientific discoveries and the beliefs he stood for.

NOTES

1   Department of Human Biology, University of Cape Town, letter from Frank Spencer to Alan Morris, 23 January 1992.
2   Department of Human Biology, University of Cape Town, Morris–Tobias correspondence, 1992–2010.

3   Department of Human Biology, University of Cape Town, letter from Spencer to Morris, 16 September 1994.

4   Tobias's graduate students often referred to him by his initials 'PVT' – Phillip Valentine Tobias. Department of Human Biology, University of Cape Town, letter from Morris to Spencer, 29 September 1994.

5   Department of Human Biology, University of Cape Town, letter from Phillip Tobias to Alan Morris, 16 September 1995.

6   Department of Human Biology, University of Cape Town, letter from Morris to Tobias, 20 October 1995.

7   Both the University of Cape Town and Wits had a programme of pure science in the first year of their respective medical courses and anatomy and physiology only began in the second year of the course.

8   'Colour Issue at UCT, Nusas President Concerned', *Cape Times*, 3 August 1948.

9   One of those holding this conservative position was Professor J.C. Middleton Shaw, the dean of dentistry, who had been recruited by Dart to begin the Dental Faculty (Murray 1997, 41).

10 University of the Witwatersrand Archives, Johannesburg, letter from Phillip Tobias to Raymond Dart, 3 June 1956.

11 Tobias maintained the friendship with many of those he met on this trip. W.W. Howells and J. Lawrence Angel would be chosen as external examiners for me when I submitted my doctoral thesis in 1984.

12 University of the Witwatersrand Archives, Johannesburg, letter from Tobias to Dart, 3 June 1956.

13 University of the Witwatersrand Archives, Johannesburg, letter from Tobias to Dart, 30 May 1956.

14 The biographical data were contributed by Ronald Singer's daughter, Hazel, for which I am most grateful.

15 Department of Human Biology, University of Cape Town, letter from Matthew Drennan to Raymond Dart, 26 June 1950.

16 Interview with Hertha de Villiers, 7 February 1990, Parkhurst, Johannesburg.

17 Interview with Ronald Singer, 20 November 1991, University of Chicago.

18 Interview, Singer.

19 Interview, Singer.

20 Interview, Singer.

21 Interview, Singer.

22 Singer commented in his interview that coloured and Indian students felt that the quality of teaching they received from the anatomists was of high value and equivalent to that white students were receiving, something also noted in more recent interviews with alumni (Perez, Ahmed and London 2012, 575).

23 Interview, De Villiers.

24 Interview, Singer.

25 Interview, Singer.

26 *Chicago Tribune*, 26 April 2006.

27 Interview with Ralph Ger, 19 March 1996, University of Cape Town.

28 See entry for Ralph Ger, 1921–2012, *Plarr's Lives of the Fellows of the Royal College of Surgeons*. https://www.rcseng.ac.uk/library-and-publications/library/blog/plarrs-lives-of-the-fellows/, accessed 9 July 2021.

29 Interview, Singer. Singer's student assistant was Kenny Abrahams who was a member of a Maoist political group in Cape Town and became a Namibian activist (Singer 1993, 107). See also https://www.klausdierks.com/Biographies/Biographies_1.htm, accessed 7 July 2021.

30 Interview, Singer.

31 Interview, Singer.

32 Iziko Museum, Cape Town, letter from Ronald Singer to Margaret Shaw, 22 November 1965; Ronald Singer, 'Physical Anthropology Research Report for 1964–65'.

33 Iziko Museum, Cape Town, letter from Singer to Shaw, 22 November 1965.

34 Interview with Alun R. Hughes, 8 February 1990, Department of Anatomy, University of the Witwatersrand, Johannesburg.

35 Curtis Marean at Arizona State received 22 boxes of Singer's files, which included much personal information, lecture preparation, some books and notes. Charles Musiba at the University of Colorado in Denver has additional uncatalogued material, which includes much of the South West Africa field notes and data and Drennan's original correspondence and notes on paedomorphosis and the 'Bushman' skull. Musiba was one of Singer's students and Marean is a researcher with a focus on southern Africa.

36 Montagu was born Israel Ehrenberg in London in 1905, but as a young man changed his name to 'Montague Francis Ashley-Montagu'. The authorship of some of his earlier publications was recorded as M.F. Ashley-Montagu, but by the 1960s he had simplified his name to Ashley Montagu.

37 Interview, Hughes.

38 Interview, Hughes.

39 Interview, De Villiers.

40 Interview, Hughes.

41 University of the Witwatersrand Archives, Johannesburg, letter from Tobias to Dart, 3 June 1956.

42 University of the Witwatersrand Archives, Johannesburg, letter from Revil Mason to Guerino Bozzoli, 12 June 1975.

43 University of the Witwatersrand Archives, Johannesburg, letter from M.H. Silk to Phillip Tobias, 18 June 1975.

44 Interview, Hughes.

45 University of the Witwatersrand Archives, Johannesburg, letter from Tobias to Dart, 30 May 1956.

46 Interview, De Villiers.

47 Interview, De Villiers.

48 Most of this information is drawn from my interview with De Villiers in February 1990, but it has been augmented from details from her application for the post of senior lecturer in the Department of Anatomy at Wits. The application of 24 June 1963 is on file in the University of the Witwatersrand Archives, Johannesburg.

49 Interview, De Villiers.
50 Baumgartel had first approached Louis Leakey, who had suggested Rosalie Osborn and then Jill Donisthorpe as field researchers. Baumgartel had hoped to fund the research with a grant from Solly Zuckerman, but Zuckerman rejected his request, saying this was an amateurish effort that did not justify financial support (Baumgartel 1976, 86).
51 University of the Witwatersrand Archives, Johannesburg, letter from Glyn Thomas to Raymond Dart, 6 February 1960.
52 University of the Witwatersrand Archives, Johannesburg, letter from Alexander Galloway to Raymond Dart, 13 February 1962.
53 Interview, De Villiers.
54 Personal communication with A.J.B. Humphreys, 31 October 2007.
55 Interview, De Villiers.
56 University of the Witwatersrand Archives, Johannesburg, 'Statement Issued by Phillip V. Tobias on His Return from Paris, 22 January 1986'.

7

Physical Anthropology and the Administration of Apartheid

In his article on the history of physical anthropology in South Africa, Phillip Tobias expressed the opinion that no South African physical anthropologist had contributed to the development of apartheid (Tobias 1985a). He was right in that no scientist with physical anthropology training had sat on the National Party committees in the late 1940s that had conjured up the complex web of legislation known as the policy of apartheid. There was no need to have a scientist on any of these committees because apartheid was never intended to be a scientific process. But Tobias was also wrong in that physical anthropologists did have other roles in relation to apartheid.

The apartheid policy implemented by the National Party codified something that already existed. South African society was already highly racialised by the end of the nineteenth century, but there was no legal structure to racial segregation. Education in the Cape Colony included elite schools for white students only (Bickford-Smith 1995, 24), but the exclusion was social rather than legislated. A partial franchise allowed educated black people to vote in the Cape Colony, but they had no such rights in the Boer republics. Miscegenation was not illegal in the Boer republics, but there was strong religious and social pressure against intermarriage between black and white people.

The real pressure for a legal system of segregation only came after the end of the South African War in 1902. Before the war, the British government had used the mistreatment and exclusion of Africans from political life as one of the excuses for the war, but it agreed to continue the exclusion of people of colour from political power in the Boer provinces of the new Union of South Africa in 1910. The Union Parliament passed the Land Act in 1913, which restricted African ownership of land, and also passed an early version of the Immorality Act in 1927, but the real stimulus for the implementation of legislated segregation came with the rise of Afrikaner nationalism in the 1930s. Saul Dubow (1995, 247) suggests that the immediate cause of apartheid (coined in 1935 and meaning separateness) was the recognition of the large degree of poverty among white Afrikaners and how they could be uplifted and protected from competition against black people. This could not work without a complex network of legislation that would define and apply formal separation.

THE CLASSIFICATION PROJECT

The policy of apartheid had been the platform of the National Party in the 1948 election. When the party won the whites-only national election in that year, it immediately began the implementation of legislation that formalised racial discrimination in South Africa. The first order of business was to stop intermarriage between people of different colours. This was accomplished with the promulgation of the Prohibition of Mixed Marriages Act (No. 55 of 1949), paired shortly thereafter with the Immorality Amendment Act (No. 21 of

1950), which reaffirmed the old 1927 Act preventing casual sexual intercourse between Europeans and 'non-Europeans'. In theory, with these two laws in place, it was now no longer legal for two people of different races to cohabit or have children. The next step was to implement formal definitions of each race and then create physical boundaries between them. This would be accomplished with the Population Registration Act (No. 30 of 1950), the Group Areas Act (No. 41 of 1950) and the Separate Amenities Act (No. 49 of 1953).

The key to everything was the Population Registration Act. This would involve the creation of a population register, where every South African resident would be recorded along with his or her specific classification. Initially, each person was to be classified into one of three categories – white, coloured or 'Bantu' – and the basis of this classification was to be physical appearance (Girvin 1987, 5). The legislators relied heavily on culture, rather than race itself, in the belief that biology and culture overlapped neatly and that a physical assignment allowed each race to be clearly compartmentalised and locked into place. 'Bantu' people were not assessed by physical appearance since membership of an African tribe and speaking a Bantu language were the cultural definitions of the race (West 1988, 103). White people were self-evident, based on their appearance and, more importantly, European culture. It was the coloured category that created the biggest problems for the apartheid administrators. Initially, the difference between whites and coloureds was exclusively based on physical appearance and discussions of descent were avoided because so many South African white people had mixed ancestry (West 1988, 109; Posel 2001, 89). The categories are presented without any attempt to justify their definitions. It is pointless to argue that the categories are irrelevant because they are arbitrary. The Population Registration Act made them uncompromisingly real.

The Population Registration Act was amended fifteen times between 1956 and 1986 (West 1988, 100), but three major revisions, in 1959, 1962 and 1967, impacted greatly on classification in general and the coloured population in particular. The 1959 amendment subdivided the original coloured category into seven separate groups: Cape Coloured, Cape Malay, Griqua, Chinese, Indian, other Asiatic and other Coloured. The category of 'other Asiatic' included such people as Zanzibar Arabs who came to Durban in the 1890s, while 'other Coloured' referred to specific groups like the many descendants of Henry Francis Fynn and his Zulu wives, who had specific historical and political rights in Zululand. Although all these people remained coloured under the Population Registration Act, the subdivisions held significance for the Group Areas and Separate Amenities acts. Up to this point, questions about identity focused on physical appearance, but by 1962 the concept of general acceptance became the most important factor in classification. If a person was accepted by the people of a particular community as a member of that community, then they would be classified according to that choice. The definitions of white and non-white were further specified and two years later were made retrospective back to 1950. The curious problem of Japanese Asians was considered in Parliament. It was decided that Japanese people should be considered white unless they applied for permanent residence, in which case they would be included as 'other Asiatic'.

The amendment of 1967 marked the last major change in the race classification system in South Africa. Since an estimated 65 per cent of the South African population had been entered on the register by this point, it now became possible to use descent as the determining factor in defining race. The classification of one's parents was all important and appearance and general acceptance were now considered secondary factors. A person was white if *both* of their parents were white; coloured if *one* or *both* of their parents were coloured, or black if *both* of their parents were black. The subdivisions of coloured were defined according to the classification of the natural father. People of

Indian origin were one of the subgroups of coloured people, but after the inception of the Tricameral Parliament in 1982 their classification gained the status of an unofficial fourth major racial/population group. Although they remained coloured in the population register, they were considered a separate category in terms of the Group Areas and Separate Amenities acts and in terms of parliamentary representation.

By 1953, 202 civil servants were employed at the task of compiling the register (Horrell 1953). A special team of investigators was created in the Bureau of Census and Statistics in order to tour the country to interview borderline or doubtful cases. Up to May 1957, some 100 000 of these borderline cases had been identified and 25 479 had been dealt with (Horrell 1957). By the end of 1966, nearly 270 000 people had been interviewed by the staff of the Bureau (Horrell 1967; see table 7.1).

Table 7.1: Classifications on the National Register at the End of 1966

Racial category	Number registered	Cases investigated as borderline	Percentage borderline of total cases
White	2 698 556	48 000	1.8
Coloured	1 107 283	193 000	17.4
Black	8 890 626	26 500	0.3
Total	12 696 465	267 500	2.1

Source: Horrell (1967)

What this meant was that slightly more than two per cent of the registered South African population had been interviewed.

The investigators responsible for the dirty work of classification were not scientists, even in the broadest definition of the term. The apartheid legislators did not believe that specialised training was necessary because ordinary people were fully capable of making racial classifications. They thought that most white South Africans understood racial differences as self-evident, but the result was that many decisions were 'capricious and arbitrary judgments' (Posel 2001, 106). The belief in the ability to classify, even when it was patently obvious that it could not be done, can be seen in the words of the minister of the interior on 15 September 1966: 'I cannot accept that there will be borderline cases for all time. If that is so, then the position is in reality so complicated that this legislation is not workable' (quoted in Horrell 1966, 123). The reality that the system was indeed unworkable would result in a cascade of objections and would draw scientists into the application – if not the administration – of the apartheid system.

RECLASSIFICATION AND THE RACE CLASSIFICATION BOARD

Not all people classified were prepared to accept the result and by the end of 1966 more than 4 000 of these individuals had lodged appeals. Inherent in the first passage of the Population Registration Act was the right of objection, and each person was given 30 days from the date of notification of classification in which to lodge an appeal. By May 1957, 1 311 objections had been received, of which 411 had been heard in court. In 1962 the route of appeal was formalised and a complainant would first request a decision from the Race Classification Board. This body had its own review board, which would consider further petitions, but the complainant also had access to the Supreme Court. A small change of significance in the appealing of a classification also came into operation in 1967. The right of a third party to object to a classification on behalf of someone else was now removed from the appeals process, but it remained possible for a third party to object *against* someone. Objections could only be lodged by the person him- or herself, a third party not on behalf of the person, or officials of the Department of Internal Affairs.

The Race Classification Board consisted of three members who sat as a court and who called evidence in review of each case. The board members themselves were not experts but sat as legal adjudicators. The chairperson of the board was a magistrate or someone with legal authority. Initially it was only the board that could make decisions about classification, but this was amended in the early 1970s to allow the director general or the secretary of the Ministry of Internal Affairs to amend a classification. The board was dissolved in 1980 and the task of reclassification was passed to the provinces to administer. Each province set itself a selection panel with a few officials who could investigate each case and make recommendations to the minister. This could include inspection of the person's place of residence, and obtaining evidence from neighbours and others.[1] These cases were all subject to appeal at the level of the Supreme Court.

The initial rush of appeals against specific classifications decreased after 1967 when descent became the primary classification method. Adding the descent rule in 1967 closed the door to people who were white in appearance but came from coloured families (West 1988, 110). At first the South African government itself did not publish any of its data, but the annual surveys of race relations in South Africa, compiled by Muriel Horrell, included comments made in Parliament during this period. Classifications were discussed only in terms of the general tripartite categories. From 1972, the annual report of the Department of Internal Affairs included census numbers, as well as the outcome of race reclassification appeals. Data about the subdivided categories of coloured people were published for the first time in this period. The richest data are from the post-1980 period. Horrell's yearly survey of race relations included the response from the minister to an annual question from liberal opposition members in Parliament. The question, asked every year, requested detailed information on the number of applications and the outcome of each application. The object of the exercise was to embarrass the government, although it is very unlikely any of the apartheid ministers saw the presentation of the numbers in a negative light since the data were

Table 7.2: Race Reclassifications in South Africa (1972–1989)

RECLASSIFICATIONS	1972	1973	1974	1975	1976	1977	1978	1979	1980	1981	1982	1983	1984	1985	1986	1987	1988	1989	TOTAL
Coloured to White	32	23	34	24	63	45	150	102	133	558	722	673	518	702	314	244	514	573	5 424
Malay to White					1					6	4	5	3		9	12	22	10	72
Chinese to White			2	2	3			1		8	7	9	7	3	7	10	4		57
Indian to White							1			1	1			1	4		4	10	22
Griqua to White																			1
White to Coloured	7	12	10	14	8	9	10	2	1	15	3	4	14	19	8	4	13	14	167
Chinese to Coloured		1	2		1			2		1				1		2	3	1	14
Indian to Coloured	13	20	26	26	23	16	10	11	4	42	34	37	50	50	63	58	62	63	608
Black to Coloured	20	22	14	17	8	17	6	1	1	79	112	88	89	249	387	269	316	369	2 064
Malay to Coloured														8	21	15	19	23	86
Griqua to Coloured															2	2			4
Malay to Chinese													1				1		2
White to Chinese					2		3	6	3	7	15	4	2			1		2	45
Coloured to Chinese	1						2	3		2	4	3	4	11	10	4	1	15	61
White to Malay							1		1	2	1				2	4		4	15
Chinese to Malay		1																	2
Indian to Malay	10	12	8	18	13	3	6	4	1	21	19	15	17	21	53	9	47	17	294
Black to Malay														3	2				5
Coloured to Malay														3	25	12	24	13	77

continued

RECLASSIFICATIONS	1972	1973	1974	1975	1976	1977	1978	1979	1980	1981	1982	1983	1984	1985	1986	1987	1988	1989	TOTAL
Malay to Indian	3	3	13	11	11	22	2	10	3	13	16	18	26	30	43	22	30	33	309
Black to Indian										6	1	2	4	2	10	5	2	13	52
White to Indian										3	1	6	1					3	16
Coloured to Indian	2	3	24	21	9	3	3	10	6	20	39	30	54	43	83	33	63	57	503
Black to Griqua										2	3	2		1	16	6	6	1	34
Coloured to Griqua															4			5	9
Coloured to Black										8	11	11	5	20	36	8	15	2	316
Griqua to Black															2				2
Indian to Black																2			2
TOTAL RECLASSIFICATIONS	88	98	134	134	144	115	198	154	153	794	993	901	795	1167	1102	722	1142	1229	10 063

Sources: Horrell (1972, 1973, 1974, 1975, 1976, 1977, 1978, 1979); Department of Internal Affairs (1980, 1981, 1982, 1983, 1984, 1985); Department of Home Affairs (1986, 1987, 1988, 1989)

also published in the annual reports of the Department of Internal Affairs (known as Home Affairs after 1986). Table 7.2 presents summarised data from the annual Department of Internal Affairs reports and Horrell's annual surveys of race relations for the period from 1972 until 1989, when race stopped being recorded in the population register.

The data tell us that more than 10 000 people changed their racial classification in South Africa in the period from 1972 to 1989. This is hardly a significant number in the scale of the total register, but it needs to be remembered that each and every one of these reclassifications was a private trauma for a person who had chosen (or had perhaps been compelled) to state their case before a government board or the Supreme Court. It is immediately apparent that the categories coloured to white (53.9 per cent) and black to coloured (20.5 per cent) make up just under 75 per cent of the total. The next largest group is the pair Indian to coloured and coloured to Indian, which together account for an additional 11 per cent of the total. Many of the remaining category changes involve fewer than 20 individuals out of the total of 10 000 plus reclassifications.

The most thorough discussion of these appeals is by Yvonne Erasmus (2007) in her doctoral thesis for the University of London. She warns that the data may be unreliable in that they are primarily from one source and therefore might be susceptible to biases, omissions, inaccuracies and inconsistencies. Geographic location, gender and age are also missing from the statistics. However, Erasmus feels that the data are the most comprehensive available and the general trends demonstrated are most likely to be true. Erasmus noted two key trends demonstrated in the data: (1) that not all the changes between the pairs of categories flowed equally in both directions; and (2) that there was an increase in the number of reclassifications after 1980.

Erasmus highlights the changes between groups within the coloured category. If the data are organised into a pairwise comparison,

Erasmus's observations become obvious (table 7.3). For the most part, the changes between coloured, Indian and Malay are roughly at the rate of 1:1. Changes between whites and Chinese and between Indians and whites are also close to 1:1. Whatever the precise motivation the individuals had for bringing their request to change to the board or court, about the same number of individuals were requesting change for each pair. Nearly all the unequal transitions are drawn from changes that involve one of the main population groups: white, Indian or coloured. In each case the differential number indicates that people were trying to reclassify into a group that had substantial social and economic advantages over the group they were currently assigned to. Being white gave someone the widest possible range of benefits (Girvin 1987, 7), but in relative terms, being coloured or Indian was an advantage over being black. Chinese identity gave a similar advantage. Although Chinese were officially part of the coloured category, in practice they did not experience a rigid separation from whites in terms of the Group Areas or Separate Amenities acts. They could not vote, but they could live in white areas and send their children to white schools if no one complained (Yap and Man 1996, 318). The Griqua were another special case. Although they were also one of the coloured subdivisions, their historical roots gave a sense of ethnic pride to people who had been relegated to an insulting mixed or bastard identity, even though much or all of their origins was from the Khoekhoe peoples of the historical Cape (Nurse 1975, 4).

Erasmus's second important observation is that a large increase in reclassifcations occurred after 1980 (2007). She notes that before 1980 there were 0.53 changes per 100 000 of the national population. From 1981, this increased to 2.81. Erasmus reports a speculation that the government wanted 'white looking coloureds' to be reclassified because they were likely to vote for the National Party in upcoming general elections (Erasmus 2007, 89). This would have been a relaxation of classification policy in response to increased pressure from

Table 7.3: Reclassifications as Pairwise Ratios (1972–1989)

Ratio Pairs	Ratio (n>10)	n	
Coloured to White:White to Coloured	32.5:1	5 591	
Black to Indian:Indian to Black	26:1	54	
Black to Coloured:Coloured to Black	17.8:1	2 180	
Black to Griqua:Griqua to Black	17:1	36	> 2:1
Malay to White:White to Malay	4.8:1	87	
Coloured to Chinese:Chinese to Coloured	4.4:1	75	
Coloured to Griqua:Griqua to Coloured	2.2:1	13	
Chinese to White:White to Chinese	1.3:1	102	
Indian to Coloured:Coloured to Indian	1.2:1	1 111	
Indian to Malay:Malay to Indian	1.0:1	603	< 2:1
Coloured to Malay:Malay to Coloured	0.9:1	163	
White to Indian:Indian to White	0.7:1	38	

Source: Compiled by the author from the data in table 7.2

anti-apartheid organisations outside of the country. But, if correct, this should have been reflected in the greater success of applicants and not an increase in numbers. There may have been a more specific reason for the increase that Erasmus did not recognise. If data are graphed over time (figure 7.1), the dramatic increase that Erasmus notes can be seen from roughly 200 cases per year to more than 1 000 cases per year after 1980. But if the data for the coloured to white and black to coloured reassignments are added, a more complicated picture emerges. The overwhelming bulk of the increase in 1981 to 1985 is from coloured to white, but from 1985 there is a second spike of reclassifications from black to coloured. What could have caused these changes and why is there a delay between the two shifts, both of which would result in a major change in social status for those reclassified?

The year 1980 was particularly turbulent in the modern history of South Africa. For the second time since 1976, students rose in protest, boycotting classes and clashing with the authorities (Welsh 2009, 281).

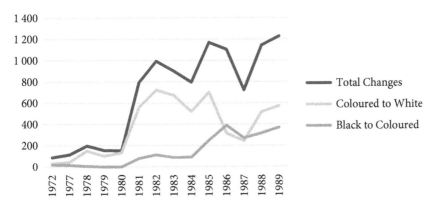

Figure 7.1: Race Classification Changes 1972–1989
Source: Compiled by the author from data in table 7.2

Initially, the boycotts were by both coloured and black students and their impact was immediate. The students called for 'Liberation now, Education later' (Mathiane 1990, 2). The education system for the next two years remained unstable, but the situation slowly improved, at least in the coloured schools. In 1983 there was a second flare-up of protests, but this time the pressure was primarily in the black schools. July and August 1984 brought another big boycott at the black schools, and this time it was extremely difficult for students to continue with their education: 'Pupils who did not wish to boycott school, or wanted to write examinations, were often threatened, and even whipped, by militants' (Welsh 2009, 287). Nearly 50 per cent of black matriculants failed between 1980 and 1989 and it was estimated that 400 000 left school with no certificate and 150 000 may have dropped out of secondary school each year in this period (Hartshorne 1992, 80). The changes in 1981 and 1985 are no coincidence. For those who were willing to expose themselves to the reclassification system, the rewards of changing from coloured to white would give their children access to the stability of the white education system. In 1981 there was no such advantage for parents changing from black to coloured, but after 1984 there was a real advantage for them to give their children access to the relatively calm learning atmosphere of the coloured schools.

How successful were the attempts at reclassification? Data on the total number of applications and the success rate are only available for the period from 1983 to 1988 (table 7.4). In most cases, if a person was prepared to go before the Race Classification Board or the Supreme Court, they were likely to win their case. The most successful reclassifications were those that changed between coloured subcategories. Changes between coloured, Malay or Indian were all reasonably successful – in excess of 90 per cent. Category changes involving black or white people were relatively more difficult to argue. In general, whites who wished to change category were not as successful as other groups wishing to change. Indians attempting to become white fared the worst.

These data on reclassification success truly show the arbitrary nature of the whole apartheid classification system. The Population Registration Act generated its own complexities. The Act limited freedom of choice and therefore invited people to attempt reclassification. But it produced separate and grossly unequal social categories, and successful reclassification could mean a new lease on life and the possibility of social, political and economic advancement.

THE SCIENTISTS GET INVOLVED

The Race Classification Board did not need scientific experts to make its decisions. In practice it carefully avoided experts and, when needed, it relied on the typological literature that abounded in the scientific journals. There could be found articles that described the skin colouration, hair form, face shape or a host of other physical features found in the different races. They were not interested in scientific argument, or frequency data or standardisation techniques. The only thing the bureaucrats on the board needed was a clear description of pure types. The literature existed and was easily accessible. The demand for

Table 7.4: Reclassification Attempts (1983–1988) where N >10

Reclassification categories	Total applications (N)	Successful applications (n)	Percentage successful
Malay to Coloured	89	86	96.6
Coloured to Malay	81	77	95.1
Coloured to Indian	388	363	93.6
Indian to Coloured	415	383	92.3
Indian to Malay	194	179	92.3
White to Malay	11	10	90.9
Malay to Indian	224	202	90.2
Coloured to Chinese	55	48	87.3
Chinese to White	41	34	82.9
Coloured to Black	117	97	82.9
Black to Griqua	35	29	82.9
White to Coloured	97	76	78.4
Black to Indian	49	38	77.6
Coloured to White	4 610	3 538	76.7
White to Chinese	12	9	75
Black to Coloured	2 457	1 767	71.9
Malay to White	89	61	68.5
Indian to White	28	19	67.9
White to Indian	15	10	66.7
TOTAL	9 007	7 026	78.0

Sources: Department of Internal Affairs (1984, 1985); Department of Home Affairs (1986, 1987, 1988, 1989)

scientific expertise came from the applicants and their lawyers and advisers. Each applicant wanted as much ammunition as possible to make their case. They hoped that the better the argument, the higher their chance would be for success.

Yvonne Erasmus and George Ellison (2008, 450) examined 69 judgements on appeals for race reclassification in the Supreme Court between 1950 and 1991. Each of these cases was an appeal against the judgement of the Race Classification Appeal Board (or its ministerial successor). The way in which evidence was presented to the board and how it was presented to the court was quite different. The Race Classification Board attempted to create compartments for each race and demanded evidence of acceptance by other members of the racial group to which the applicant was requesting transfer, while the court took a more balanced view of each appellant's life. Judges in the Supreme Court were reluctant to get involved in scientific evidence, but the applicants frequently enlisted scientists to help them to make their case. In interviews with two scientists who were involved with this process, Erasmus and Ellison note that the scientists co-operated because they wanted 'to do the best they could for the people concerned' (Erasmus and Ellison 2008, 451). Both scientists were extremely uncomfortable about their involvement, but felt their involvement was not to co-operate with the system but 'to help somebody whose life was going to be devastated by what was a fiendish and unjust law' (Erasmus 2007, 211). These two scientists did not provide evidence before the board itself because they felt that classification could not be done with any form of reliability and the 'scientific tests' were most often based on pseudoscience.

So, who were the scientists that Erasmus interviewed and whose cases were described in Erasmus and Ellison (2008)? Erasmus has chosen not to identify her sources because her interviews were based on anonymity, but she has confirmed that neither of her scientific sources was trained specifically as a physical anthropologist.[2] Both appeared only once for a race classification case at the Supreme Court, but one of her experts indicated that he had provided expert opinion for more than twenty clients. This makes sense in light of the testimony of physical anthropologists (see below). Although on occasion expert testimony was required in court or before the board, for the most part the intersection of science and society occurred at the level of legal opinion. The background of Erasmus's two scientists is unknown, but anthropological scientists were in the public eye because of their discoveries. The geneticist Trefor Jenkins found that requests for assistance with racial classification came frequently after the 1963 *Man in Africa* exhibition in the Johannesburg city hall, which included a display on human variation (Kuljian 2016, 149). Many physical anthropologists were happy to give interviews to newspapers on various topics of interest, so it was not difficult for lawyers or other interested parties to find them. Many of these contacts were informal or did not require a written report. We know of some of them because of old correspondence stored in archival files, but other cases were described in interviews. What is obvious from the collected reports is that the intersection between science and apartheid race classification was more common than Erasmus has described in her research. However, there are no common patterns as each case presented unique circumstances.

The need to affirm classification occurred even before apartheid regulations took hold. Raymond Dart received a request for help in June 1947 from the Witwatersrand Central School Board.[3] In this case, the parents of a boy from Blantyre in Nyasaland (modern-day Malawi) had applied for him to come to a European-only school in Johannesburg. The parents were Greek and German, but the child was 'very dark and gives indications of a non-European descent'. The letter from the Witwatersrand Central School Board explains that the doubt 'appears to be on the mother's side'. The secretary of the school board requested that Dart examine the boy and furnish the board with a confidential report for what was 'a particularly difficult case'. Sadly, we do not have a record of Dart's response.

For Ted Keen, the exposure to race classification came not from the Race Classification Act itself, but from the Immorality Act. In the early 1950s, Keen was approached by the family of a young man who had been charged with picking up a prostitute and was now doubly in trouble because the woman was a coloured person. The young man argued that he had no idea the woman was coloured, but the state argued that she was identifiable as coloured, even when she was wearing make-up. Keen was asked in his capacity as an anatomist to visit the woman where she was being held in prison, and to assess whether it would be difficult for the young man to identify her as a person of colour. Although Keen did examine the woman, he argued that no one could tell the difference at night and that he 'didn't see why this chap should be sent to jail for breaking your laws'.[4] Keen's case is typical of the conundrum created by petty apartheid. Whatever his thoughts about collaborating with the apartheid structures, the real issue was that lives were being disrupted and destroyed by these laws and getting involved in the legal process could help individuals to deal with the very real immediate problem.

Christa Kuljian (2016, 152–153) tells the very personal story of how Hertha de Villiers dealt with a request for classification of a baby girl in 1966. In March of that year, a small delegation, consisting of a social worker from the child welfare society along with a nurse and a baby from the Princess Alice Adoption Home, visited Phillip Tobias at the Medical School. They requested help in the classification of the baby, so that they could make a formal decision about putting her up for adoption as there was a question about the racial origins of the father. Such requests were not uncommon when there was a question about the racial incompatibility of the parents. Adoption from the Lady Buxton Children's Emergency Home in Claremont, Cape Town, was often delayed for as much as a year after birth until the racial features of the baby were confirmed.[5] Often the final assessment was done by a bureaucrat from Pretoria, who would arrive with a checklist of features (hair form, nose shape, presence of a Mongol spot at the base of the spine) to examine the child. In the 1966 case, Tobias was uncertain and asked De Villiers, then a lecturer in the Department of Anatomy, to assist him. De Villiers was moved both by the baby and by the uncertainty of what would happen to the child if she were classified coloured. She concluded that the baby's features fell 'definitely within the range of variation of the features encountered in Europeans of European national origin' and then adopted the baby herself (Kuljian 2016, 152).

These difficult decisions became even more complex when the anthropologists were before the Race Classification Board itself. Tobias remembered one case that he was directly involved in.[6] Sometime in the mid-1960s, he was called to present evidence in a case under consideration by the board. A wealthy businessman with a white wife was brought before the board on the assumption that he was Indian. The government was pressing the case and if he were identified as Indian he would be prosecuted under the Mixed Marriages Act. Tobias said he was a 'Caucasoid' individual whose features could fit well with populations from southern or eastern Europe. The man was declared white, but Tobias was never asked to testify before the board again.

Although the Race Classification Board was not a court of law, lawyers were often deeply involved in the process of reclassification. In February 1966, a legal firm in Johannesburg approached Lawrence Wells and Trevor Trevor-Jones to provide an affidavit on the race of their client. They met the man concerned and offered a non-committal report, indicating that he could not 'be described either as obviously a white person or as obviously not a white person'.[7] Wells provided a very detailed three-page explanatory note on race classification along with his affidavit. The board rejected this ambiguous report and classified the client as non-white. The client and his lawyers challenged this in the Supreme Court on 20 April 1967 and both Trevor-Jones and Wells were called as expert witnesses. Unfortunately, we do not have a record of the outcome of the Supreme Court proceedings, so we do not know if the client succeeded in being reclassified white.

Trefor Jenkins and George Nurse of the Department of Human Genetics at the University of the Witwatersrand (Wits) were called in a similar way to provide evidence from a serogenetic profile before the Race Classification Board.[8] A woman from Botswana wished to be called white because she wanted to come to South Africa to marry a white man. Jenkins and Nurse reported to the board that nothing appeared in her genetic spectrum to suggest she was not white, but the board rejected their evidence because the people in her family did not look white. Nurse and Jenkins wrote to the newspapers saying that the minister was disregarding science. In response, the minister wrote to them to tell them that race was political, not a scientific issue, and like Tobias, neither Nurse nor Jenkins was asked to testify again.

For the most part, the scientists who were engaged in these very personal questions of racial appearance chose to resist the system by arguing that the applicant should be classified into the new category of their choice. There was clearly a motive for the physical anthropologists to engage with the process in the hope that they could be of service to the individuals involved. But at least two scientists took the view that even engaging with the process was a moral issue. Although Ronald Singer left no record of his contact with the system of race classification, he told his American colleagues that he had been asked to be involved in a race case, but had refused.[9] In the mid-1980s I was approached by a law firm asking for help in a case being brought before the Race Classification Board. In this case, a man had been running a general goods dealership in Khayelitsha township in Cape Town, but the place had been looted and burned on two occasions during unrest. His insurance company had notified him that it was no longer prepared to cover him, and the man knew that his only option was to move his business elsewhere or to suffer destitution if his premises were attacked again. But the Group Areas Act limited him to African areas. Although he was accepted as a black person and his home language was isiXhosa, he argued that he was the son of an Indian man and a black African woman and that he had been raised by his mother. On this basis he now wanted to change his classification to Indian so that he could move his business to an Indian area, but he had no proof of his father's identity and could not provide the Race Classification Board with the necessary documents. The lawyer wanted me to meet his client, assess the racial identity of his physical features (in a similar way that Dart, Tobias and Wells had done) and go to the board to testify that this man was Indian. Unlike Dart, Tobias and Wells, I chose not to co-operate. I was not prepared to deal with an apartheid legal structure and the request went against my scientific understanding of human variation and the arbitrary nature of racial classification.

The refusal of myself and Singer to give advice in these cases must not be viewed as a higher moral judgement in comparison to the co-operation of our colleagues. Although I do not know the outcome of the case, my decision not to get involved may have cost this man his livelihood. What our refusal does do is to highlight the dilemma that each of these scientists faced when they considered their science in relation to the society around them. For Dart and Wells, this may not have been a central issue since they were convinced that science was objective and did not present an issue of right or wrong, but dealing with the inequalities of race in South Africa made Tobias, Jenkins and myself uncomfortable and forced us to refine our own definitions of right and wrong.

LEARNING NEW ROPES: TYPOLOGY AND ITS CONSEQUENCES

The implementation of apartheid after 1948 was a political process that was out of step with most of the world after the Second World War. The advent of the 'new physical anthropology' in the 1950s influenced the ideas of Tobias and others as the decade progressed, but it had no influence on the structure of the apartheid legislation as it was applied by the government. The ideology that supported apartheid came out of the central European theory of ethnos as imported to South Africa by the Afrikaans-speaking *volkekundists* in their own version of ethnology (Sharp 1981, 19) and it was supported by the pre-existing tenets of typology and diffusionism. Although the local physical anthropologists did create novel ideas of their own and formed a kind of South African scientific identity (especially for the Afrikaans-speaking scientists), it was the European baggage of ethnos, typology and diffusionism that coloured much of the South African research.

The balance of evidence suggests that physical anthropologists were not at the forefront of the design and implementation of the policy of apartheid, but they did involve themselves with the process of race reclassification at the request of individuals undergoing this process. Perhaps a more insidious guilt can be laid at the feet of some of them in the promotion of an unquestioning strict vision of typology. Typology is not racism, but its inflexibility and inability to deal with ranges of variation provide fertile soil in which racism can be nurtured. This was not unique to South Africa, but we are a special case because of the complexity of our population history, our colonial roots and the social engineering of the apartheid government. Tobias's claim that I alluded to at the start of this chapter might be technically true, 'but it neglects the extent to which interwar physical anthropology both contributed to and expressed the broader climate of scientific racism in South Africa' (Dubow 1995, 28).

The age of typology coincided with the rise of eugenics and the South Africanisation of science where practical knowledge in the social and natural sciences was controlled by a white elite (Dubow 2006). Unlike the United States, Great Britain and Germany, the eugenics movement in South Africa was never a strong organisation that influenced legislation, but its ideas did help others to form their opinions (Dubow 2010, 279). One of the reasons why eugenics did not gain a more prominent position in South African society was that it clashed with some of the ideals of Afrikaner nationalism. Eugenicists believed feeble-mindedness, seen mostly in poorly educated communities, should be controlled with a programme of sterilisation. This was unpopular among Afrikaner nationalists because Afrikaans speakers made up most of the poor white community. Where eugenics and Afrikaner nationalism did overlap was in the public fear of racial mixing. Eugenics equated miscegenation with loss of racial fitness, a fear that was prominent in the awareness of the lay public. *Volkekunde*, the academic arm of Afrikaner nationalism, was concerned with the separation of peoples into their pure types, hence its focus on coloured research (Walters 2018, 96). These academics incorporated a negative view of racial mixing from eugenics and their racial stereotypes were generated by the physical anthropologists.

Dart tolerated the extremes of scientific racism in his belief that science could be impartial. He refused to acknowledge the implications of his typology and diffusionism. As his more outlandish ideas slowly crumbled in the scientific realm, they were maintained in the eyes of the lay public because they fitted what the public wanted to believe. From a scientific perspective, it was Tobias who 'maintained the image and reputation of Dart despite the idiosyncrasies of so much of his work', even though it was also Tobias's work that overturned many of Dart's conclusions, including those on race (Derricourt 2010, 233). Robin Derricourt argues that the image of Dart would have faded or been addressed more critically without Tobias's intervention. Dubow (1996, 27) concurs, noting how Tobias sought to play down the shift in emphasis between pre-war and post-war physical anthropology, downplaying the implicit racism in the Dartian tradition.

A lesser question is whether any of the earlier researchers were explicitly racist and supported the ideals of apartheid when it began. With the exceptions of Coert Grobbelaar and Con de Villiers at Stellenbosch, who were actively involved in the teaching of German national socialist models of physical anthropology, and Egbert van Hoepen in Bloemfontein, who was explicitly against the advancement of black peoples, there is no evidence that any of the physical anthropologists were outright racists. Dart and Wells tried to avoid any form of political opinion despite both appearing to have essentially liberal values. Wells expressed hope in correspondence with Dart asking for a letter of reference for the chair of anatomy at Cardiff University in the 1960s. He was sad at the thought of leaving South Africa: 'It is a difficult decision. In spite of present discontents there is a lot to be said in favour of life in South Africa, and we still do not see the problems of this country as insoluble.'[10]

Matthew Drennan was the only one who expressed a positive opinion of apartheid. He had a special interest in the colonial history of South Africa. Writing to the *Cape Times* in 1937, Drennan lamented the damage being done to the old Supreme Court building by the widening of the Wale and Adderley street intersection and argued that poor urban planning was damaging economically valuable historical monuments.[11] He was concerned about the preservation of the seventeenth-century hedge that marked the edge of the Dutch colonial lands along the eastern side of Table Mountain, and he was captivated with the traces of European culture that had been transplanted to the Cape Colony by both the Dutch and the Huguenot settlers.[12] Drennan's interest in the Cape Dutch may have been spurred by his contact with Afrikaans-speaking acquaintances in the Cape's health administration. His links to the provincial health services gave him social perks, such as invitations to social occasions of government departments, and he became very good friends with several National Party dignitaries.[13]

Drennan's interest in prehistory and his connections with the Cape government led to him being appointed to the board of the South African Museum in 1955. The boards of the museums were strongly influenced by politics in the 1940s and 1950s. The chair of the boards of the South African Museum, the South African Art Society and the Parliamentary Scientific Society was one man, Senator David van Zyl. He saw the museums as places to showcase Afrikaner ideals and actively supported the selection of 'ideologically motivated men' to the boards of the museums, so that they could support the apartheid dream (Mazel 2013, 167). Under Van Zyl's influence, a programme was launched to create a South African Cultural History Museum to showcase white South African and European material and cultural history. Aron Mazel (2013, 170) argues that Drennan was seen as politically conservative and supportive. An English-speaking member of the board of the South African Museum was rare under Van Zyl's chairmanship, and many of the other members were strongly linked to the racist ideals of the National Party. Among Drennan's fellow board members was Con de Villiers, the overtly racist professor of zoology from the University of Stellenbosch. About a month before his death in 1965, Drennan wrote to Van Zyl expressing his appreciation of the latter's work in support of museums in South Africa and labelled museum directors who had left as political deserters (Mazel 2013, 172).

Drennan was broadly supportive of the apartheid legislation that was unfolding around him in the 1950s, not as a fanatical racist, but more as a colonial remnant who saw Europeans as the rightful carriers of higher civilisation. Writing from a personal perspective to his old school in Edinburgh, Drennan lauded the Edinburgh medical model that had been used for the creation of the University of Cape Town (UCT) Medical School, but also noted that the South African experience differed from the Scottish because of the complexity of the racial elements in South Africa (Drennan 1951, 8). After reviewing the historic peopling of the subcontinent, Drennan remarked how European South Africans had taken on the parental responsibility to guide the economic, educational and social future of black and coloured South Africans. His conclusion noted how 'South Africa of to-day is simply refusing to be overwhelmed by exotic biological forces and ideologies, but she is willing and anxious to work towards a holistic and harmonious internal federation of diverse peoples and cultures' (Drennan

1951, 11). It would be difficult to present a stronger statement in support of the policy of apartheid, but in many ways Drennan was expressing the views of thousands of English-speaking white South Africans. The National Party had stoked the fears of white South Africans by equating the African liberation movements with international communism and declaring that giving power to black people would be the end of civilisation in southern Africa. The National Party used a similar trick in the 1982 general election by convincing the electorate that giving separate Houses of Parliament to Indian and coloured South Africans would draw these two groups of people into the apartheid fold in support of whites against blacks.

The introduction of apartheid came at a time when the view of anthropology was changing. The United Nations Educational, Scientific and Cultural Organization (Unesco) *Statement on Race*, published in 1950 (Spencer 1997c), was interpreted differently by different audiences. The *Statement's* claim that race was a sociological and not a biological construct was interpreted as meaning there was no such thing as biological race. The architects of the *Statement* intended to express the view that the concept was not valid, but that human variation still existed and was something that should be studied at the finer level of population. But many of the so-called progressive groups, including the Black Consciousness organisations, interpreted this literally as meaning that since race does not exist, human variation was not a significant characteristic of the human species. For many in these organisations, the study of physical anthropology itself was of the old order and should not be part of modern anthropology. Writing in 1986 to *Sechaba*, the official organ of the African National Congress (ANC), the exiled activist Archibald Crail talked about how physical anthropologists at UCT identified culturally pure Khoekhoe by hair texture, facial characteristics, body build and so on. Certainly by 1986 no such research was being done or published from UCT, but Crail wrote as if it was still going on. There was a need to ensure that anger against the inequities of apartheid continued, even if that required the need to falsely state that race scientists of the old order still existed (Morris 2009a, 79). The Cape Action League, a Cape Town activist group aligned to the Pan Africanist Congress (PAC), produced a workshop reader to train community activists on the issue of race and racism. The document identified two kinds of progressive scientists: the splitters, led by Tobias, who continued to study discrete groups of people who could be distinguished by inherited characteristics; and the lumpers, led by Ashley Montagu, who taught that the human species could not 'be neatly or even approximately' divided into racial groups (Cape Action League 1987, 5). It then went on to present a strict typological division of human variation into four races as an example of the splitter position, arguing that this was Tobias's viewpoint. The authors emphasised that 'we have a vested interest in ensuring that the lumpers position becomes universally accepted' (Cape Action League 1987, 7). To demonstrate the rejected splitter position, the workshop reader included a clinal map (presumably of skin colour), neatly labelled with a legend identifying five races (one more than mentioned in the text). Unfortunately, they got it wrong, and the authors of the reader understood neither Montagu nor Tobias. Clinal maps, as discussed in Tobias's work, among others, are very much part of the lumpers' argument, indicating the difficulty in delineating typological races, and are never shown as indicating races – only continuous distributions of traits. Martin Hall (1990, 73), referring to the failure of Black Consciousness organisations to understand the modern approaches to Late Iron Age archaeology, raises the fact that many of the people in these organisations were the product of poor-quality apartheid education. Not only did they distrust white academics, but archaeological scholarship tended to be technical and unintelligible to non-academics. The same could be said about physical anthropology.

In the twenty-first century, physical anthropology, pulling its baggage of unwelcome history, faces a real challenge on two fronts: the need to understand how the science of the observation of human variation has changed, and how to explain that to a wider audience. We cannot simply ignore our unpleasant roots. Instead, we must use the lessons learned to find a pathway that honours both scientific knowledge and its meaning in greater society.

NOTES

1   Interview with Attie Tredoux, Chief Legal Officer of the Department of Home Affairs from 1978 to 1999, 16 November 1994, Cape Town.
2   Letter from Yvonne Erasmus to author, 22 May 2020.
3   University of the Witwatersrand Archives, Johannesburg, letter from the Witwatersrand Central School Board to Raymond Dart, 17 June 1947.
4   Interview with Ted Keen, 6 December 1994, Department of Anatomy, University of Cape Town.
5   Personal communication from Liz Morris in 1980. She was then working as a nurse at the Lady Buxton Home.
6   Interview with Phillip Tobias, 7 May 1992, Department of Anatomical Sciences, University of the Witwatersrand.
7   University of the Witwatersrand Archives, Johannesburg, letter from Lawrence Wells to G.W. Petersen, 10 February 1966.
8   Interview with George Nurse, 2 July 1992, Department of Anatomical Sciences, University of the Witwatersrand.
9   'Ronald Singer', *Chicago Tribune*, 26 April 2006.
10 University of the Witwatersrand Archives, Johannesburg, Lawrence Wells to Raymond Dart, 16 November 1966.
11 'Defence of Old Supreme Court Building', *Cape Times*, 29 September 1937.
12 Department of Human Biology, University of Cape Town, Matthew Drennan, 'Transplanting Traditions', 1946.
13 Interview with Ronald Singer, 20 November 1991, University of Chicago.

8

The Politics of Racial Classification in Modern South Africa

The typological construction of biological races appeared to many to be an attractive methodology because it explained so much human variation without having to worry about confusing complexities, but it is those complexities that are the reality. This is the problem that confronted the practitioners of the 'new physical anthropology' as they tried to move away from typology. During the 1960s and 1970s two techniques were added to the armament of physical anthropology analysis. Both had their roots in the first half of the twentieth century.

OF SKULLS, BLOOD AND STATISTICAL ANALYSIS

Mention has already been made of Karl Pearson's biometric school and its mathematical approach to quantifying humanity. Pearson created something he called the 'coefficient of racial likeness', which was a mathematical formula that added up the differences in a series of measurements of two crania or two samples (Pearson 1926). The idea was interesting but ultimately unworkable because the measurements were intercorrelated (they measured similar things) and this made any conclusions unrealistic.

The solution to this problem was found by comparing the variation seen in the measurements, rather than looking at their averages. Hertha de Villiers (1968) chose to use Lionel Penrose's method (Penrose 1952), which separated out size variation and gave a better calculation of shape, while G. Phillip Rightmire (1970) and I (Morris 1984) worked with more complex multivariate techniques, which were only manageable with the advent of computers. None of these techniques identified race. What they did was to provide an accurate mathematical description (termed 'biological distance') of how crania differ in appearance. The most recent methods use geometric morphometrics. Images of bones are collected with a laser scanner and mathematical distance is computed from the three-dimensional spread of previously defined points – all done on a computer screen without having to touch a measurement tool to bone. This has proved invaluable for the detailed description of shape and is now commonly used in comparing differences between species where morphology has functional or adaptive importance.

But are these not just better methods of creating types? This is certainly a fair question as the techniques can indeed be used in exactly the same manner as the old typological measurements. For the most part, multivariate statistical analyses are being used in bioarchaeology research, which is interested in relationships between local populations, and race is not even part of the discussion. In 1996 two anthropologists working at the University of Tennessee, Stephen Ousley and Richard Jantz, produced a discriminant function software package called Fordisc, which could be used to identify race from cranial measurements in forensic investigations. A set of standard measurements are taken on an unknown skull and the program compares these measurements to a set of reference skulls. The calculation places the unknown skull as close as possible to the best-fitting group. The Fordisc programme works reasonably well, but it is limited by the fact that the reference samples do not reflect all human cranial variations. The discriminant function will force a skull into the nearest best-fit group and this can sometimes lead to spectacular errors (Morris 2010b, 18). Ousley and Jantz also introduced an arithmetic tweak to the program called a sectioning point. This effectively draws a line on the graph that allows the researcher to make a clear-cut decision as to which group the unknown skull falls into. Although based on probabilities and methodologically correct, the use of a sectioning point downplays the reality of overlap between the racial subdivisions, especially because discriminant analysis accentuates differences between individuals in its attempt to find a best fit.

Why would anthropologists who have studied human variation and understand the difficulty in fitting racial types to reality still be prepared to use this mathematical application? The explanation can be found in legal practice, not science. The problem for the courts is that scientific evidence must be clear and precise to be acceptable (Morris 2011, 51). The testimony of an expert witness that is not based on well-defined scientific method will be contested on the basis of it being opinion rather than fact. This is a worrying issue because scientific evidence can sometimes be inappropriate or based on a false premise. The court may be accepting evidence on the underlying assumption that science is always right. The reference samples to which the unknown forensic case is compared will have a biological meaning to the scientist, but this may be very different from racial identity perceived by the police and understood in the courts (Morris 2010b, 29).

Multivariate statistical analyses of cranial shape have been useful for archaeology (and arguably for forensic anthropology), but a very different approach developed in the 1960s to look at relationships between living people. This was serogenetics, or the analysis of genetic markers from blood samples. The presence of genetic polymorphisms (different expressions of the same gene) in red blood cell antigen markers, enzymes and serum proteins had been discovered early in the twentieth century. Initially, these characters were used in a similar fashion to cranial characters and were listed as identifiers of racial types (Pijper 1930, 311; Elsdon-Dew 1935, 653; Grobbelaar 1955, 323), but as the understanding of genetics improved the technique became the analysis of choice for the identification of a new form of biological distance, one based on the genes, rather than outward appearance. The serogeneticists argued that since polymorphisms in blood were controlled by genes within the cell nucleus, the expression of specific blood markers was a direct reflection of the underlying genotype, or genetic blueprint. The serogenetic technique was indeed better than the earlier methodology, but it was not genetically neutral and was distorted by the effects of natural selection on the frequencies of the blood polymorphisms (Morris 2005c, 112). High (or low) frequencies of particular markers could be a sign of shared genetic history, but might also be the result of natural selection by similar environmental pressures such as disease events.

In the same manner as craniological measurements, the frequency of blood polymorphisms could be manipulated by statistical techniques and a hierarchy of populations could be generated according to similarity in trait proportions in the populations. This led to an opportunity to check whether the serogenetics data confirm the major races of the old typologists. It did not (Weiss and Maruyama 1976, 31), but it did lead to a very interesting observation. There was a close congruence in regional patterns between craniology, serological markers and language (Howells 1989; Cavalli-Sforza, Menozzi and Piazza 1994). The markers, whether osteological or serogenetic, were tracking local conditions of adaptation, isolation or population movement. What they were really good at was telling us about the life experiences of populations, but not about their distant origins.

Scientists have struggled to interpret the patterns seen in the geographic data. George Nurse, Joseph Weiner and Trefor Jenkins published a compilation of two decades of their work on the peoples of southern Africa titled *The Peoples of Southern Africa and Their Affinities* (1985). They avoided the old morphological measurements and focused on the serogenetics data. My own negative review in *Anthropology Today* (Morris 1988a, 3) triggered a detailed response from Nurse and one from Vernon Reynolds, another physical anthropologist with an interest in environmental variations in humans. My complaint about the book was that the authors chose a very simplistic view of South African history and prehistory. Based on an assumption that Khoesan populations originated in desert climates and 'Negro' populations originated in tropical forests, they chose to link these climates to physical traits of morphology in exactly the same way as the old typologists had done. Not only did archaeology confirm that their climatic origin assumptions were wrong, but their simple interpretations of complex anatomical features were jarring. Nurse, Weiner and Jenkins also proposed an old theory of prior adaptation to explain that each population had produced its own special constellation of evolved traits, which either consciously or unconsciously impelled a population to search for an environment similar to the one it had left, explaining human migrations as biological compulsions, rather than historical events. Both Nurse's and Reynolds's responses were spirited and provided the meat for a lively debate, but both stood by the claims for physical adaptations to broad environmental conditions and motivations for migratory movements of peoples (Nurse 1988, 24; Reynolds 1988, 25). Both correspondents accused me of taking a political rather than a scientific line, something reminiscent of the complaints against Ashley Montagu in his *Statement on Race*, published in 1951. My book review and these letters provided such a lively debate that the editors of the *World Archaeological Bulletin* decided to reprint the originals verbatim.

SCIENTISTS RE-ENGAGE WITH THE PROBLEM OF RACE CLASSIFICATION

Even with its faults, serogenetics did provide a good method to compare populations, but did it really put the nail into the coffin of the old racial system of classification? The most influential paper from the serogenetics stable was that of Richard Lewontin in 1972. He gathered a large, worldwide series of serum protein and red blood cell enzyme variants and analysed his sample at three levels: as a single group of individuals, as clusters of individuals representing local populations and as five major races. His results indicated that 85.4 per cent of variation could be found between any two individuals and that only 8.3 per cent of additional variation could be found between different populations. Most startling of all was that the differences between continental races accounted for just 6.3 per cent of total variation (Lewontin 1972, 383). The results indicated that human variation is overwhelmingly at the level of individuals and biological differences between groups are too complex to be explained by races (Caspari 2010, 52).

Lewontin's 1972 paper on blood proteins has been used to argue that biological race does not exist. Interpretations of his research have suggested that genetic differences between people are so trivial that biological traits can be ignored and that any research into race is pseudoscientific. The reality is that Lewontin's work demonstrates why race is a bad way to describe geographic variation, but that does not mean that geographic variation is non-existent or unimportant. The major flaw in his study is the belief that the blood polymorphisms accurately represent the genetic differences between individuals and the assumption that the number of differences could simply be added together to indicate differences between people. This is known as a phenetic analysis, in which all measurements/observations are of equal statistical weight (Hull 1988, 158). In reality, the impact of natural selection means that different serogenetic features may have very different reasons for having been selected by one group or another, and rarer blood markers (and their underlying genes) might have greater importance in one population compared to another. Simply put, not all genes are equal.

The era of serogenetics ended abruptly with the development of molecular genetics, the study of the structure of the deoxyribonucleic acid (DNA) molecule itself. Both the craniologists and the serogeneticists had been searching for a way to assess true biological distance between any two people. DNA analysis allows examination of the genetic code directly, rather than through an osteological feature or blood protein proxy. For the first time, a technique had been discovered that offered a map of genetic relationships between living peoples and provided an objective measure of distance between them (Morris 2005b, 112). Genetic maps were made of the hypervariable regions of mitochondrial DNA and the Y-chromosome. A conscious effort was made to avoid points on the chromosome that were known to code for specific characteristics. Fortunately, there is a great deal of redundancy in the DNA molecule and there are non-coding parts in which mutations would accumulate without the impact of natural selection. The number of different mutations in these DNA regions between any two individuals will reflect the length of time since they last shared a common ancestor. These are used to track lineage (female and male lines of inheritance). The most recent development in molecular genetics has been the ability to do whole genome analyses in which it is possible to examine the subset of genes that identifies the multiple lines of inheritance that reflected the complexity of population origin. Whole genome analysis has also allowed us to look at the coding portions of the DNA in order to see the effects of natural selection, especially the segregation of medically important genetic strains in separate populations (Reich 2018).

The first attempts to compare real genetic variation between large geographic races (Africa, Asia and Europe) brought a surprising result. All analyses of African populations showed them to be far more variable than the similar-sized populations in Europe or Asia (Cann, Stoneking and Wilson 1987, 31; Vigilant et al. 1991, 1503). More differences mean more mutations accumulated, which means more time since they shared a common ancestor with Europe and Asia. Southern Africa has held a special place in this research because of all of the world's populations, it is the Khoesan of southern Africa who harbour the greatest genetic diversity (Schlebusch, Lombard and Soodyall 2013). This places them near the root of modern humans (Schlebusch et al. 2017, 652). Please do not think this makes them some sort of remnant primitive in the manner of Raymond Dart. The Khoesan of southern Africa have had a long and complicated history, which we are only now beginning to understand (Barbieri et al. 2014; Chan et al. 2015), but they are just as modern as any other living human population.

What molecular DNA research has provided us with is a method to understand the complexity of human variation, without having to use outmoded explanations like racial types. The geographical patterns seen in craniological, serological and linguistic studies start to make sense in two ways. Natural selection through differential mortality, chance selection of rare genotypes by isolation and restrictive breeding patterns, and movement of genes through migration and interbreeding between populations have shaped humanity at the level of the local population throughout history and prehistory. This is what the older methodologies have been showing us and the whole genome analysis has confirmed, but DNA has also allowed us to track lineages back through time. This has shown us the commonality of humanity.

David Reich has been at the forefront of whole genome research into ancient DNA and the origins of modern populations (Reich 2019). Among his conclusions is that the old typological myth of pure races has no support in genetics. Migration and mixture are part of human history and living populations are all the result of multiple ancestors, both biologically and culturally. Reich emphasises that there are real and significant genetic differences between populations, but that even if two populations can be shown to have different average values for traits (and therefore their underlying genes), individuals in the group are not defined by the average. The difference between groups is real, but it is overwhelmed by differences between individuals. In the same manner that we recognise that men and women differ physically and genetically from each other, those differences are coloured by social expectations and upbringing (Reich 2018).

The struggle to understand the difference between the use of race as a diagnostic technique of separating people and the very real patterns of worldwide geographic variation continues. In May 2007 the American Association of Physical Anthropologists held a symposium on how human variation is seen by practitioners of the subject (Edgar and Hunley 2009, 1). Lineage tracking by DNA markers has clearly shown modern humans to be a subset derived from African variants (Hunley, Healy and Long 2009, 45) and the observed pattern of genetic variation does not support a simple three-group racial categorisation, but there remains a considerable difference of opinion about the quantification of geographic variation. John Relethford (2009, 16) has demonstrated that the worldwide geographic patterning in skin colour and cranial form produces observable regional differentiation, that cranial shape is a good reflection of regional variability and that statistical classification techniques do have high success rates. This is the paradox faced by forensic anthropology practitioners using forensic discrimination programs such as Fordisc. It is obvious that people from different parts of the world look different. The problem lies not in the presence of these variations, but in how they are quantified.

Physical anthropologists have not been very successful in explaining how their view of race has changed from the bad old days of typology and race science. Much of this rests in the technicality of the subject. It seems trite to say that large groups of people (races) are meaningless, but smaller groups (populations) are different. Is this not just about numbers? George Chaplin and Nina Jablonski (2020, 134) have argued that only species have real taxonomic meaning and that all subdivisions of humanity are arbitrary. They note that very few human populations are truly isolated. The human species as a whole is one gene pool that is constantly changing and biogeographic entities (such as races) are temporary. Eloquent though their argument is, it makes no sense. Species are a product of evolution, not evolutionary units in themselves. Evolutionary adaptation cannot happen in vast populations as there is too much complexity for selection. Although it is true that few human populations are genetic isolates, all human populations have restrictions of gene flow between groups. Chaplin and Jablonski are correct in that geographic human groupings (demes) are of little importance because of the migratory behaviour of modern humans, but restrictions in gene flow are the result of cultural choices in marriage partners, not geographic distance.

The irony for humans is that genetic (Mendelian) populations have meaning because they are culturally defined and cultural choices have biological consequences. Genetic drift and founder effect, both aspects of statistical chance, are genetic processes that happen in small groups, but are spread to larger groups by gene flow. Large racial groups do not have real biological meaning as their complex genetics do not represent natural selection on a large scale. These large-scale groupings (races) reflect the complex history of the many small populations from which they originated and therefore are historical, rather than taxonomic. Crain Soudien (2020, 117) has tried to take issue with Reich about this, but it is more about terminology than fact. The two are not speaking the same language. Where Reich sees genetically distinct populations, Soudien sees arbitrary racial categories. Soudien refers to Reich as a doubter who has not accepted the consensus that race does not exist, yet what Reich is saying is that there are indeed real genetic differences between groups of people, at both large and small scales. Working in the field of ancient DNA, Reich has been tracking these genetic differences back through time to see how they produced what we have today – one continuously variable interbreeding species made up of multiple small populations, each of which continues to undergo migration, drift and adaptation at the local level.

CONFLICTING VIEWS ON RACE IN POST-1994 SOUTH AFRICA

While the new physical anthropology has struggled with how to best explain the difference between the old concept of race and the modern understanding of human variation, the South African lay public is still confronted with racial categories that remain as real as they were in the dark days of apartheid. It would have been wonderful to be able to say that racial discrimination and inequality ended with the demise of apartheid, but the legacy of apartheid has remained, and the focus has now shifted to finding ways to redress past imbalances. Although the political landscape has changed, the majority of South Africans still live in segregated communities and socio-economic status continues to be stubbornly stratified by racial origin.

As the end of apartheid approached in 1991, the administrators of apartheid made the decision to stop recording race as a category on the population register. At least on paper, for three years until the first open election in 1994, South Africans lived in a society in which race was not assigned. But as the new democracy dawned, it became obvious that righting the wrongs of the past required the ability to know who had been disadvantaged so that redress could be offered to them. Although the brand-new Constitution professed to aspire to non-racialism, legislation was passed to recognise previously disadvantaged people, especially in terms of economics. The term 'race' was avoided and instead reference was made to population group. Statistics South Africa allowed individuals on the census to self-classify using a four-scale system – black, coloured, Indian/Asian, white – and a similar approach was used in the creation of quotas in a range of situations from university entrance to the selection of sports teams. Employers asked their employees to define themselves as how they were (or would have been) classified before the 1994 elections. The Broad-Based Black Economic Empowerment Act of 2003 defined black people – Africans, Coloureds, Indians and, shortly afterwards, Chinese – as designated groups. Unlike the apartheid legislation, no definitions of the groups were provided and the categories relied on the 'common sense of the existence of apartheid racial groups' (Stone and Erasmus 2012, 134). The irony, of course, is that by using race as a proxy for disadvantage, the legislators were running the risk of causing injustice and intolerance while they were transforming society. Redress legislation created the paradox in which post-1994 legislation continued 'to contribute to the existence and perpetuation of the concept of race by creating the impression that these categories are natural and valid' (Stone and Erasmus 2012, 139).

A similar debate about the use of racial classification took place in the pages of the *South African Medical Journal* (*SAMJ*). For the legislators, race was a proxy for social ills and a history of disadvantage, but clinical medicine was also concerned about the purely biological issues of how disease presented itself and was treated in the individual. In the same manner as the physical anthropologists, the clinicians were confronted with separating ethnicity (sociocultural aspects of human groups) from the genetically based biological characters of people. George Ellison et al. (1996, 1257) ponder the definition and use of race, ethnicity and population group in public health research and recommend that none of them be used in segregating data. In their opinion, none of these terms are helpful in finding the root cause of inequality in health and medical treatment and they create a situation where disadvantaged individuals who do not fall into the appropriate category would be missed in health interventions. For Ellison and his colleagues, public health is about socio-economic disadvantage that is not bounded by racial/ethnic/population group categories. These terms perpetuate the process of discrimination and do not identify the root cause of disadvantage.

This debate was of particular importance to the various editors of the *SAMJ* as it was up to them not only to provide guidance to authors, but also to make firm decisions about whether the research would be published. Daniel Ncayiyana (2007, 1225) recognised that when citizens are asked to state their race in official medical documents, they are doing this in the context of 'a society that seeks to erase "race" consciousness and stereotyping'. Ncayiyana and his editorial staff on the *SAMJ* had already decided on a policy not to publish papers that made unwarranted reference to race, but were confronted by epidemiologists who wanted to continue to use race-based analyses to press for post-apartheid redress. Ncayiyana realised that although race was being used as a surrogate for the causes of ill-health, the concept was poorly suited for health determinants that were not genetically based. Ncayiyana did not deny that ethnicity (shared culture, values, language, descent, religion and so on) is real and as important to health as biological differences, but he asked researchers to be careful in deciding what is to be measured and not to confuse social and economic variables with genetic determinants.

Ellison et al. (1996, 1257) and Ncayiyana (2007, 1225) would have chosen not to use any racial or ethnic terminology at all, if that were possible, but Anton van Niekerk (2011, 248) took a more pragmatic position in support of linking health issues to racial categories. Van Niekerk asked whether the benefits in terms of diagnosis and research outweighed the social cost of using race as a factor in the selection of a study group. When race is identified as a marker, the research protocol for each study has to carefully consider whether all other possible variables have been considered. Simply removing race as a variable in research would be unwise as 'it represents tensions and contestations that will continue to permeate interpersonal contact and social relations in this country' (Van Niekerk 2011, 250).

A small but important point raised by Van Niekerk is whether the ethnic or population group names chosen by the researcher are acceptable to the groups themselves. Jacquie Greenberg wrote to J.P. Van Niekerk, the editor of the *SAMJ*, to ask if the journal had a specific policy on how to standardise group nomenclature.[1] She was concerned about the naming of local populations defined by race or ethnicity from which blood samples had been drawn for genetic analysis. The journal did not have such a policy, so Janet Seggie, the new editor, requested opinions on the matter from colleagues. I reported on a symposium held in June 1971, the conclusions of which were later summarised by two of the attendees (Jenkins and Tobias 1977). The symposium had aimed to create a uniform and consistent set of names for the peoples of southern Africa to be used by natural, biomedical and social scientists.[2] Although 40 years out of date, the symposium did provide one very useful suggestion. It had noted that terminology had to be specific to the set of variables that were being considered. One set of names would refer to biology, a second set to language and a third to economy or culture. Biology and culture were neither fixed nor linked and classifications could and would change over time. I summarised my own recommendations to anatomists (Morris 2003, 2), in which I strongly recommended that the human sample chosen should first be thought of in terms of community, then ethnicity and only lastly biology. Care should be taken that information from small samples is not scaled up to generalise about large groups since medical evidence drawn from local populations may not be true for arbitrary races. Ncayiyana wrote from his position as a previous editor of the journal.[3] He said that no formal policy had been in place in the past and that he expected each author to explain the relevance and rationale for invoking any given classification. He recommended that researchers be aware of self-defined names of people and urged them to use these names where possible and he reiterated my warning about using small data sets to generalise about large groups consisting of thousands (if not millions) of people.

The use of self-defined names is an important concept that is not considered often enough by scientists. Although scientists profess to be arbitrators in the definition of biological populations and ethnic groups, very few understand even the basic difference between the meaning of race and the meaning of ethnicity (Štrkalj and Wilkinson 2006, 112). Names matter because they form a link between the researcher and those being studied. Katharina Schramm (2016, 138) has highlighted how archaeological skeletons referred to as 'specimens' by researchers are spoken of as 'ancestors' by the descendant communities. There is an immense gap between the perceptions of scientists, who appoint themselves as knowers and the indigenous peoples as knowledge objects, and the perceptions of indigenous peoples. The relationship is one-sided and unequal from the perspective of the descendant communities. The conflict in perception also holds true in medical research when scientists are insensitive to the self-identity of the communities they are visiting.

Schramm has used my attempt to build forensic facial reconstructions of the historic eighteenth-century burials from Cobern Street in Cape Town as an example of this problem (Schramm 2020, 344). My aim was well intentioned in that I believed that presenting reconstructed faces instead of dry bones would help in the dissemination of the results of the study of the Cobern Street skeletons to the wider public. I thought that providing close visual approximations of appearance in life might help descendant communities to understand the value of the research. The facial reconstructions were completed, but the project itself was never published or used as part of an outreach to local people who claimed descent from those buried at the cemetery. Among the problems encountered was that the process of reconstructing the faces was not free of racial identity. Only about two-thirds of the facial features are based on the measured attributes of the underlying bone (Wilkinson 2010, 235) and as a result the forensic odontologist and artist who completed the faces drew heavily on assumptions of race to produce the final lifelike images (Morris, Phillips and Rosendorff 2012). Even though the five crania had been selected to represent three distinct cultural traditions seen in the burial practices, these had been interpreted by the artist as representing racial groups for the purposes of placing the fine-scale facial features on the reconstructions. As Schramm notes, the key problem is that the public antagonism towards anything that uses racial classification is still so powerful that any attempt to explain the procedure of reconstruction would be rejected outright as part of the problem, rather than part of the solution. From Schramm's perspective, the anomaly is that both sides try to reject the concept of race, yet race remains in the eyes of both.

The two perspectives, one from biology and one from sociology, are hard to reconcile. Elsewhere, I have suggested reasons why many social scientists still maintain that the study of physical anthropology is racial science (Morris 2005a, 18). There is no question that the subject is tainted by its own past. Racial origin was nearly the only subject of interest for much of the history of the field. Raymond Dart, Matthew Drennan and others focused on classifying groups, rather than understanding why they might or might not be different. Typological views of race are still strong in the public forum, although they are disappearing from the scientific one. The lay public hears typological comments about race in a wide range of media, but they rarely hear the scientific refutation of such comments. Physical anthropology has failed to actively counter racist conclusions and as a result has allowed the idea to fester that nothing has changed in its scholarship. This lack of educational outreach has given validation to an assumption that scientists still think that the only valuable thing to study in physical anthropology is race. Although it is true that physical anthropology no longer considers large geographic races as appropriate units for study, it is a myth that human variation is not a valid subject of study. But this is not just about how non-scientists fail to understand current scholarship in the field. Schramm's lesson is a warning that it is also the scientists themselves who fail to understand the sensitivity and anger that exists among communities who have suffered under racial oppression.

The insensitivity of scientists to the opinions of the lay public has shown itself in very specific ways. Late in 2019 a paper on the cognitive functioning of South African coloured women was published by a team of five University of Stellenbosch academics (Nieuwoudt et al. 2020). Based on a very small sample of just 60 individuals, the researchers concluded that coloured women, as a racial group, were at risk of low cognitive function because they had low educational levels and unhealthy lifestyles. The criticism was immediate and so intense that the journal, *Aging, Neuropsychology, and Cognition*, decided to retract the article despite the fact that it had already been published online. Part of the criticism was because of the poor science that allowed such a sweeping generalisation from very limited data, but other criticism was directed at how such research was accepted by the ethics committee of the University of Stellenbosch and how the published version had passed peer review. Steven Robins (2020, 43) was concerned that such research demonstrated the racial, ethnic, religious and cultural stereotyping that was present outside of the academic context. Lesley le Grange (2019, 11) takes broader issue with the whole of modern Western science, which she conflates with the

Eurocentric vision of the natural world, and argues that there can be no such thing as neutral science because it is always culturally and historically bounded. She goes on to note that the issue is about 'who controls the means and processes of knowledge production and who exercises power over subjectivity, and indirectly to power over economy, given the current age of academic capitalism and the ascendency of neoliberal politics, policies and universities' (Le Grange 2019, 13). I am not sure if I can support such a generalisation about Western science, but Le Grange does have some strong points. Sharné Nieuwoudt et al. (2020) mixed poor research design and execution with a near total failure to understand the meaning of groups of people. Their research is divorced from the social and political context of science in South Africa and has added to the negative view in which social science researchers such as Le Grange perceive biomedical researchers.

This conflict of perception has most recently manifested in a paper by Nicoli Nattrass (2020) on why so few black African students choose biology as a subject at university level. The paper raised a significant uproar both for its subject matter and for the way it posed its research question. Essentially its question was raised in such a way that it appeared that the problem was the black students' inability to appreciate science as a career in comparison to careers that had a greater chance of monetary reward. The university itself was not considered as the source of the low number of biology registrations. In a stinging critique in an online newspaper, Ferial Haffajee (2020) pointed out that South Africa has a whole generation of black professionals who steward its national parks, but few of them were drawn from an academic pathway in the universities.

The editors of the *South African Journal of Science* saw this debate as so important that they produced a special journal issue just to present the criticisms of Nattrass and her response. Chief among these critics was Jimi Adesina (2020, 5), who demonstrated how the underlying assumptions made the conclusions unviable. Nattrass had drawn her data from a once-off survey of science students at the University of Cape Town (UCT), but this sample did not reflect the broader picture of science students at other universities, nor did it consider the impact of long years of underfunded postgraduate study in an educational context very different from the students' home lives. Many of the other commentaries raised similar points about research methodology and underlying assumptions, but two papers in particular raised points of relevance to racial classification and how it is perceived within and outside the university. Jeremy Midgley (2020, 22), a fellow UCT academic, noted the pressure placed on departments to consider the race of their students, with the explicit demand to increase the intake of black African students at both undergraduate and postgraduate level. Essentially this creates an institution where racial generalisations become the proxy for individual student needs. Lindelani Mnguni (2020, 25) takes the racial debate from another perspective by linking it to the current #BlackLivesMatter discussion. He blames the entire conflict on institutional racism, which assumes that black students are outsiders who need to adapt to the institution, rather than the other way round. Mnguni sees problems in methodology as extensions of colonised education. In the same way that Schramm saw the debate about the Cobern faces as a product of different views of race, the extreme polarisation in the debate about the Nattrass paper is very much a product of academics failing to understand the viewpoints of others. The background of the researcher (or her critics) cannot be separated from the science under discussion.

Where Nattrass was absolutely correct is that the perception of students is strongly influenced by both cultural and socio-economic contexts in exactly the same manner as the academics who teach them. Lars Guenther and Peter Weingart (2016, 3) found that young educated South Africans see science and technology as a route to progress and that it can aid in the development of the country, but many fear that science will change their lives too much. Too much science can challenge beliefs and faith and can risk the loss of their cultural and social roots. Lower socio-economic groups with lower levels of education showed this reservation to an even larger extent, making scientific ideas even more difficult for them to accept. Scientific claims (such as human evolution) that conflicted with cultural beliefs were believed the least. Guenther and Weingart did not raise the issue of race or racial classification, but it is easy to see that these topics would generate the same sort of conflict in the minds of the lay public as the scientific claims about human origins.

Heather Edgar and Keith Hunley (2009, 4) lament that American students still hold strong typological views of race because anthropologists have failed to offer clear and satisfactory explanations that mesh with the life experience of students. It is hard to explain that race is not a viable concept in the light of very real physical differences seen in people from different regions of the globe. The historian Nancy Stepan has best described this in explaining why the lay public in Britain continues to accept simplistic typological explanations of race:

A science that requires considerable genetical and mathematical knowledge to grasp, and whose essence is populational and statistical, is not one that is easily translated into non-scientific language and absorbed into public discourse. The 'new' science of human diversity also seems to be denying the 'obvious' – is counter-intuitive. After all, everyone 'knows that races do exist'! Typological thinking is far easier than statistical thinking, and the continued existence of racial tensions in society has provided many opportunities for the expression of old, typological racial prejudices. (Stepan 1982, 182)

For most South Africans race is as relevant today as it was during the height of apartheid. The legacy of our racist past still clings to our society in the form of inequality of groups in terms of economic advantage and this will not change until these inequalities are gone. Scientists must claim part of this negative historical legacy and they must be active in explaining not only how their subject has changed, but also how it became part of the problem.

NOTES

1   Copy in possession of the author of a letter from Jacquie Greenberg to J.P. van Niekerk, 23 January 2013.
2   Letter from Alan Morris to Janet Seggie, 19 February 2013.
3   Copy in possession of the author of a letter from Daniel Ncayiyana to Janet Seggie, 3 March 2013.

SELECT BIBLIOGRAPHY

NEWSPAPERS

Cape Argus
Cape Times
Eastern Province Herald
The Friend
The Star
The Friend

ARCHIVES

Albany Museum, Grahamstown
American Museum of Natural History Archives, New York
Ditsong National Museum of Natural History, Pretoria
Iziko Museum, Cape Town
McGregor Museum, Kimberley
National Museum, Bloemfontein
Natural History Museum, London
Port Elizabeth Museum, Gqeberha
Royal College of Surgeons of England Archives, London
University of Aberdeen, Aberdeen, Scotland
University of Cape Town, Department of Human Biology, Cape Town
University of Cape Town Special Collections, Cape Town
University of South Africa (Unisa) Institutional Repository (online at http://uir.
    unisa.ac.za, accessed 16 July 2021)
University of the Witwatersrand Archives, Johannesburg

INTERVIEWS

Marie Barry (née Dreyer), 29 May 1993, Oudtshoorn
Hertha de Villiers, 7 February 1990, Parkhurst, Johannesburg

Baz Edmeades (Thomas Dreyer's grandson), December 1992, Cape Town
Ralph Ger, 19 March 1996, University of Cape Town
Alun R. Hughes, 8 February 1990, Department of Anatomy, University of the
    Witwatersrand, Johannesburg
Ted Keen, 6 December 1994, Department of Anatomy, University of Cape Town
Ralph Kirsch, 5 May 1992, Department of Anatomy, University of Cape Town
George Nurse, 2 July 1992, Department of Anatomical Sciences, University of the
    Witwatersrand, Johannesburg
Margaret Shaw, 26 June 1991, South African Museum, Cape Town
Ronald Singer, 20 November 1991, University of Chicago
Phillip Tobias, 7 May 1992, Department of Anatomical Sciences, University of the
    Witwatersrand, Johannesburg
Attie Tredoux, chief legal officer of the Department of Home Affairs from 1978 to
    1999, 16 November 1994, Cape Town

PUBLICATIONS AND PRESENTATIONS

Adesina, Jimi O. 2020. 'The Anatomy of a Bad Science: Reflections on Nattrass' Commentary'. *South African Journal of Science* 116: 5–12. https://doi.org/10.17159/sajs.2020/8523.

Badenhorst, André Riaan. 1968. 'Egbert Cornelis Nicolaas van Hoepen'. In *Dictionary of South African Biography, Volume 4*, edited by W.J. de Kock, 700–701. Pretoria: National Council for Social Research, Department of Higher Education.

Balsan, François. 1954. *Capricorn Road*. London: Arco Publications.

Barbieri, Chiara, Mario Vicente, Sandra Oliveira, Koen Bostoen, Jorge Rocha, Mark Stoneking and Brigitte Pakendorf. 2014. 'Migration and Interaction in a Contact Zone: mtDNA Variation among Bantu-Speakers in Southern Africa'. *PLOS One* 9, no. 6. https://doi.org/10.1371/journal.pone.0099117.

Basu, Paul and Vinita Damodaran. 2015. 'Colonial Histories of Heritage: Legislative Migrations and the Politics of Preservation'. *Past and Present* 226, supplement 10: 240–271. https://doi.org/10.1093/pastj/gtu028.

Baumgartel, Walter. 1976. *Up among the Mountain Gorillas*. London: Robert Hale.

Bickford-Smith, Vivian. 1995. *Ethnic Pride and Racial Prejudice in Victorian Cape Town*. Johannesburg: Wits University Press.

Binneman, J. 1990. 'A History of Archaeology in the Eastern Cape: Part 1, The Spade Brigade'. *The Phoenix* 3, no. 1: 12–15.

Bolk, Lodewijk [Louis]. 1929. 'Origin of Racial Characteristics in Man'. *American Journal of Physical Anthropology* 13, no. 1: 1–28. https://doi.org/10.1002/ajpa.1330130123.

Bolwig, Niels. 1988. *From Mosquitoes to Elephants: A Life with Animals*. Cheltenham: Self-published.

Booth, Clive. 1988. 'James Drury (1875–1962)'. *Sagittarius* 3, no. 1: 22–24.

Brink, Vernon H. 1923. 'A Preliminary Genetic Study on the Osteology of the Griquas'. *Transactions of the Royal Society of South Africa* 11, no. 1: 145–169. https://doi.org/10.1080/00359192309519576.

Brink, Vernon H. 1933. 'A Genetic Study of the Vertebral Column of the Griqua'. *Annals of the University of Stellenbosch* 11, no. 1 (Series A). Cape Town: Nasionale Pers.

Broom, Robert. 1918. 'The Evidence Afforded by the Boskop Skull of a New Species of Primitive Man (*Homo Capensis*)'. In *Anthropological Papers of the American Museum of Natural History* 23, part 2: 63–79. New York: Published by order of the Trustees [of the American Museum of Natural History].

Broom, Robert. 1923. 'A Contribution to the Craniology of the Yellow-Skinned Races of South Africa'. *Journal of the Royal Anthropological Institute* 53: 132–149.

Broom, Robert. 1929. 'Australoid Element in the Koranas'. *Nature* 124: 507. https://doi.org/10.1038/124507a0.

Broom, Robert. 1941. 'Bushmen, Koranas and Hottentots'. *Annals of the Transvaal Museum* 20: 217–249.

Brown, Alec C. 2003. 'Centennial History of the Zoology Department, University of Cape Town, 1903–2003: A Personal Memoir'. *Transactions of the Royal Society of South Africa* 58, no. 1: 11–34. https://doi.org/10.1080/00359190309519932.

Burns, Catherine. 1996. 'Bantu Gynaecology: The Science of Women in South Africa, 1920–1956'. Paper presented at a Centre for African Studies Seminar, University of Cape Town, 8 May.

Burrows, Edmund H. 1958. *A History of Medicine in South Africa.* Cape Town: Balkema.

Camp, C.L. 1948. 'University of California African Expedition: Southern Section'. *Science* 108, no. 2812: 550–552. https://doi.org/10.1126/science.108.2812.550.

Cann, Rebecca L., Mark Stoneking and Allan C. Wilson. 1987. 'Mitochondrial DNA and Human Evolution'. *Nature* 325, no. 6099: 31–36. https://doi.org/10.1038/325031a0.

Cape Action League. 1987. *Introduction to 'Race' and Racism: Workbook 1.* Cape Town: Cape Action League.

Carstens, Peter, Gerald Klinghardt and Martin West. 1987. *Trails in the Thirstland: The Anthropological Field Diaries of Winifred Hoernlé.* Centre for African Studies Communications No. 14. Cape Town: Centre for African Studies, University of Cape Town.

Caspari, Rachel. 2010. 'Race and Geographic Variation Conflated: An Impediment to Teaching Human Biology'. In *Teaching Human Variation: Issues, Trends and Challenges*, edited by Goran Štrkalj, 43–59. Hauppauge, NY: Nova Science Publishing.

Caton-Thompson, Gertrude. 1983. *Mixed Memoirs.* Gateshead: Paradigm Press.

Cavalli-Sforza, Luigi L., Paolo Menozzi and Alberto Piazza. 1994. *The History and Geography of Human Genes.* Princeton: Princeton University Press.

Chan, Eva K.F., Rae-Anne Hardie, Desiree C. Petersen, Karen Beeson, Riana M.S. Bornman, Andrew B. Smith and Vanessa M. Hayes. 2015. 'Revised Timeline and Distribution of the Earliest Diverged Human Maternal Lineages in Southern Africa'. *PLOS One* 10, no. 3. https://doi.org/10.1371/journal.pone.0121223.

Chaplin, George and Nina Jablonski. 2020. 'Semantics in the Philosophy of Race'. In *Persistence of Race*, edited by Nina G. Jablonski, 128–139. Stellenbosch: African Sun Media.

Chiarelli, Brunetto and Giuseppe D'Amore. 1997. 'Italy'. In *The Encyclopaedia of the History of Physical Anthropology*, edited by Frank Spencer, 529–536. New York: Garland.

Comrie, John D. 1932. *History of Scottish Medicine*. 2nd edition. London: Baillière, Tindall & Cox.

Correia, Janine C., Quenton Wessels and Willie Vorster. 2013. 'Edinburgh, the Scottish Pioneers of Anatomy, and Their Lasting Influence in South Africa'. *Scottish Medical Journal* 58, no. 4: 256–260. https://doi.org/10.1177%2F0036933013507839.

Crail, Archibald. 1986. 'Letter to the Editor'. *Sechaba* (July): 30–32. Accessed 30 July 2021. https://disa.ukzn.ac.za/sites/default/files/pdf_files/sejul86.pdf.

Cunningham, Andrew. 2010. *The Anatomist Anatomis'd: An Experimental Discipline in Enlightenment Europe*. Farnham: Ashgate.

Dahlberg, Gunnar. 1942. *Race, Reason and Rubbish*. Translated by Lancelot Hogben. London: George Allen and Unwin.

Dart, Raymond A. 1923. 'Boskop Remains from the South East African Coast'. *Nature* 112: 623–625. https://doi.org/10.1038/112623a0.

Dart, Raymond A. 1925a. '*Australopithecus Africanus*, the Man-Ape of South Africa'. *Nature* 115: 195–199. https://doi.org/10.1038/115195a0.

Dart, Raymond A. 1925b. 'The Present Position of Anthropology in South Africa'. *South African Journal of Science* 22: 73–80. Accessed 30 July 2021. https://hdl.handle.net/10520/AJA00382353_8557.

Dart, Raymond A. 1929a. 'A Note on the Taungs Skull'. *South African Journal of Science* 26: 648–658. Accessed 30 July 2021. https://hdl.handle.net/10520/AJA00382353_4034.

Dart, Raymond A. 1929b. 'Phallic Objects in Southern Africa'. *South African Journal of Science* 25: 553–562. Accessed 30 July 2021. https://hdl.handle.net/10520/AJA00382353_4028.

Dart, Raymond A. 1929c. 'The South African Negro'. *American Journal of Physical Anthropology* 13, no. 2: 309–318. https://doi.org/10.1002/ajpa.1330130228.

Dart, Raymond A. 1930. 'With the Italian Scientific Expedition'. *The Headlight* (December): 24–25.

Dart, Raymond A. 1937a. 'The Physical Characters of the /?Auni-‡Khomani Bushmen'. *Bantu Studies* 11: 175–246. https://doi.org/10.1080/02561751.1937.9676051.

Dart, Raymond A. 1937b. 'Racial Origins'. In *The Bantu-Speaking Tribes of South Africa*, edited by Isaac Schapera, 1–32. Cape Town: Maskew Miller.

Dart, Raymond A. 1939. 'Population Fluctuation over 7000 Years in Egypt'. *Transactions of the Royal Society of South Africa* 27: 95–145. https://doi.org/10.1080/00359193909519788.

Dart, Raymond A. 1940. 'Recent Discoveries Bearing on Human History in Southern Africa'. *Journal of the Royal Anthropological Institute of Great Britain and Ireland* 70, no. 1: 13–27. https://doi.org/10.2307/2844198.

Dart, Raymond A. 1960. 'In Memoriam: Matthew Robertson Drennan'. *South African Medical Journal* 39, no. 36: 828–829.

Dart, Raymond A. 1966. 'A Tribute to Reginald Ruggles Gates, FRS'. *Mankind Quarterly* 7, no. 1: 48–52.

Dart, Raymond A. 1972. 'Associations with and Impressions of Sir Grafton Elliot Smith'. *Mankind* 8: 171–175. https://doi.org/10.1111/j.1835-9310.1972.tb00431.x.

Dart, Raymond A. 1974. 'Cultural Diffusion from, in and to Africa'. In *Grafton Elliot Smith: The Man and His Work*, edited by A.P. Elkin and N.W.G. Macintosh, 160–174. Sydney: Sydney University Press.

Dart, Raymond A. and Dennis Craig. 1959. *Adventures with the Missing Link*. London: Hamish Hamilton.

Dart, Raymond A. and Nino del Grande. 1931. 'Ancient Iron-Smelting Cavern at Mumbwa'. *Transactions of the Royal Society of South Africa* 19, no. 4: 379–427.

Davison, Patricia. 1993. 'Human Subjects as Museum Objects: A Project to Make Life-Casts of "Bushmen" and "Hottentots", 1907–1924'. *Annals of the South African Museum* 102, no. 5: 165–183.

De Almeida, António. 1956. 'La Macronymphie chez Les Femmes Indigènes de l'Angola'. *Comptes Rendus de l'Association des Anatomistes*, 43a Reunião, 131–150. Lisbon, 26–29 March.

De Kok, J.M. 1987. 'Coert Smit Grobbelaar'. In *Dictionary of South African Biography, Volume 5*, edited by C.J. Beyer and J.L. Basson, 307. Pretoria: Human Sciences Research Council.

De Villiers, Hertha. 1963. 'A Biometrical and Morphological Study of the Skull of the South African Bantu-Speaking Negro'. PhD diss., University of the Witwatersrand.

De Villiers, Hertha. 1968. *The Skull of the South African Negro: A Biometrical and Morphological Study*. Johannesburg: Wits University Press.

De Zwart, Pim. 2011. 'South African Living Standards in Global Perspective, 1835–1910'. *Economic History of Developing Regions* 26, no. 1: 49–74. https://doi.org/1 0.1080/20780389.2011.583003.

Deacon, Hilary J. and Janette Deacon. 1999. *Human Beginnings in South Africa: Uncovering the Secrets of the Stone Age*. Cape Town: David Philip.

Deacon, Janette. 1979. 'Guide to Archaeological Sites in the Southern Cape'. Guidebook prepared for the Southern African Association of Archaeologists Excursion, 30 June–4 July.

Deacon, Janette. 1990. 'Weaving the Fabric of Stone Age Research in Southern Africa'. In *A History of African Archaeology*, edited by Peter Robertshaw, 39–58. London: James Currey.

Dechow, Paul C. 2006. 'In Memoriam, Ronald Singer, 1924–2006'. *The Anatomical Record* 289B: 114–115. https://doi.org/10.1002/ar.b.20109.

Department of Home Affairs. 1986. *Annual Report: 1 July 1984 to 30 June 1985*. Pretoria: Department of Home Affairs.

Department of Home Affairs. 1987. *Annual Report: 1 July 1985 to 30 June 1986*. Pretoria: Department of Home Affairs.

Department of Home Affairs. 1988. *Annual Report: 1 July 1986 to 31 December 1987*. Pretoria: Department of Home Affairs.

Department of Home Affairs. 1989. *Annual Report: 1 January 1987 to 31 December 1988*. Pretoria: Department of Home Affairs.

Department of Internal Affairs. 1980. *Annual Report: 1 July 1979 to 30 June 1980*. Pretoria: Department of Internal Affairs.

Department of Internal Affairs. 1981. *Annual Report: 1 July 1980 to 30 June 1981*. Pretoria: Department of Internal Affairs.

Department of Internal Affairs. 1982. *Annual Report: 1 July 1981 to 30 June 1982*. Pretoria: Department of Internal Affairs.

Department of Internal Affairs. 1983. *Annual Report: 1 July 1982 to 30 June 1983*. Pretoria: Department of Internal Affairs.

Department of Internal Affairs. 1984. *Annual Report 1 July 1983 to 30 June 1984*. Pretoria: Department of Internal Affairs.

Department of Internal Affairs. 1985. *Annual Report: 1 July 1984 to 30 June 1985*. Pretoria: Department of Internal Affairs.

Derricourt, Robin. 2009. 'The Enigma of Raymond Dart'. *International Journal of Historical Studies* 42, no. 2: 257–282.

Derricourt, Robin. 2010. 'Raymond Dart and the Danger of Mentors'. *Antiquity* 84: 230–235. https://doi.org/10.1017/S0003598X00099890.

Desmond, Adrian. 1989. *The Politics of Evolution: Morphology, Medicine and Reform in Radical London*. Chicago: University of Chicago Press.

Digby, Anne. 2013. 'Black Doctors and Discrimination under South Africa's Apartheid Regime'. *Medical History* 57, no. 2: 269–290. https://dx.doi.org/10.1017%2Fmdh.2012.106.

Drennan, Matthew R. 1913. *Application for the Lectureship in Anatomy in the South African College, Cape Town*. Edinburgh: Darien Press.

Drennan, Matthew R. 1921. 'Some Problems in Medical Science Awaiting Solution in South Africa'. *South African Medical Record* 19, no. 15: 283–287.

Drennan, Matthew R. 1922a. 'Embryology: Its Scope and Aims'. *South African Medical Record* 20, no. 7: 134–135.

Drennan, Matthew R. 1922b. *A Short Course on Human Embryology*. Cape Town: Citadel Press.

Drennan, Matthew R. 1925. 'Was South Africa "the Cradle of Mankind"?' *Illustrated London News*, 5 September: 432.

Drennan, Matthew R. 1927a. 'A Contribution to the Piltdown Problem'. *Nature* 120: 874. https://doi.org/10.1038/120874a0.

Drennan, Matthew R. 1927b. 'The Homologies of the Arm and Leg'. *Anatomical Record* 35, no. 2: 113–128. https://doi.org/10.1002/ar.1090350207.

Drennan, Matthew R. 1927c. *Illustrations for a Short Course on Human Embryology*. Privately published.

Drennan, Matthew R. 1929a. 'An Australoid Skull from the Cape Flats'. *Journal of the Royal Anthropological Institute* 59: 417–427. https://doi.org/10.2307/2843893.

Drennan, Matthew R. 1929b. 'The Dentition of a Bushman Tribe'. *Annals of the South African Museum* 24, no. 1: 61–87.

Drennan, Matthew R. 1930a. *A Short Course in Physical Anthropology*. 2nd edition. Cape Town: Mercantile Press.

Drennan, Matthew R. 1930b. 'A South African Human Skull Cap'. *South African Journal of Science* 27: 547–548. Accessed 30 July 2021. https://hdl.handle.net/10520/AJA00382353_3580.

Drennan, Matthew R. 1931. 'Pedomorphism in the Pre-Bushman Skull'. *American Journal of Physical Anthropology* 16: 203–210. https://doi.org/10.1002/ajpa.1330160214.

Drennan, Matthew R. 1937. 'Some Evidence for a Trepanation Cult in the Bushman Race'. *South African Medical Journal* 11: 183–191. Accessed 30 July 2021. https://hdl.handle.net/10520/AJA20785135_12395.

Drennan, Matthew R. 1938a. 'Archaeology of the Oakhurst Shelter, George: Part III – The Cave-Dwellers'. *Transactions of the Royal Society of South Africa* 25, no. 3: 259–280. https://doi.org/10.1080/00359193709519753.

Drennan, Matthew R. 1938b. 'Archaeology of the Oakhurst Shelter, George: Part IV – The Children of the Oakhurst Cave-Dwellers'. *Transactions of the Royal Society of South Africa* 25, no. 3: 281–293. https://doi.org/10.1080/00359193709519754.

Drennan, Matthew R. 1949. 'The Dentition of the Cape Coloured as Shown by a Study of 150 Male and Female Coloured Skulls'. *South African Dental Journal*: 1–10.

Drennan, Matthew R. 1951. 'A "Synapse" with the University of Cape Town'. *Synapse: Edinburgh Medical Student's Journal* 3, no. 3: 7–11.

Drennan, Matthew R. 1953a. 'A New Race of Prehistoric Men: The Saldanha Skull'. *Illustrated London News*, 26 September: 480–483.

Drennan, Matthew R. 1953b. 'A Preliminary Note on the Saldanha Skull'. *South African Journal of Science* 50, no. 1: 7–11. Accessed 30 July 2021. https://journals.co.za/doi/abs/10.10520/AJA00382353_2514.

Drennan, Matthew R. 1963. 'The Early College Years'. *South African Medical Journal* 37, no. 3: 52–53. https://www.ajol.info/index.php/samj/article/view/180137.

Drennan, Matthew R. and Ronald Singer. 1955. 'A Mandibular Fragment, Probably of the Saldanha Skull'. *Nature* 175: 364–365. https://doi.org/10.1038/175364a0.

Dreyer, Jacobus. 1988. 'Die Hoffman-Angola-Ekspedisie'. *Culna* 35: 8–9.

Dreyer, Thomas F. 1931. 'The Bushman-Hottentot-Strandlooper Tangle'. *Transactions of the Royal Society of South Africa* 20, no. 1: 79–92. https://doi.org/10.1080/00359193109518851.

Dreyer, Thomas F. and Albert J.D. Meiring. 1937. 'A Preliminary Report on an Expedition to Collect Old Hottentot Skulls'. *Soölogie Navorsinge van die Nasionale Museum, Bloemfontein* 1, no. 7: 81–88.

Dreyer, Thomas F. and Albert J.D. Meiring. 1952. 'The Hottentot'. *Researches of the National Museum, Bloemfontein* 1, no. 1: 19–22. Accessed 30 July 2021. https://hdl.handle.net/10520/AJA00679208_11.

Dreyer, Thomas F., Albert J.D. Meiring and Abraham C. Hoffman. 1938. 'A Comparison of the Boskop with Other Abnormal Skulls from South Africa'. *Zeitschrift für Rassenkunde* 7: 289–296.

Drury, James and Matthew R. Drennan. 1926. 'The Pudendal Parts of the South African Bush Race'. *Medical Journal of South Africa* 22, no. 4: 113–117.

Du Pisani, Etienne. 1989. 'Anthropological and Archaeological Pursuits at the National Museum, Bloemfontein'. Paper presented at the 53rd South African Medical Association Conference, Bloemfontein, April.

Dubow, Saul. 1995. *Illicit Union: Scientific Racism in Modern South Africa*. Cambridge: Cambridge University Press.

Dubow, Saul. 1996. 'Human Origins, Race Typology and the Other Raymond Dart'. *African Studies* 55: 1–30. https://doi.org/10.1080/00020189608707838.

Dubow, Saul. 2000. 'A Commonwealth of Science: The British Association in South Africa 1905 and 1929'. In *Science and Society in Southern Africa: Studies in*

Imperialism, edited by Saul Dubow, 66–99. Manchester: Manchester University Press.

Dubow, Saul. 2006. *A Commonwealth of Knowledge: Science, Sensibility and White South Africa, 1820–2000*. Oxford: Oxford University Press.

Dubow, Saul. 2010. 'South Africa: Paradoxes in the Place of Race'. In *The Oxford Handbook of the History of Eugenics*, edited by Alison Bashford and Philippa Levine, 274–288. Oxford: Oxford University Press.

Eales, Hugh. 2007. *Riddles in Stone: Controversies, Theories and Myths about Southern Africa's Geological Past*. Johannesburg: Wits University Press.

Edgar, Heather J.H. and Keith L. Hunley. 2009. 'Race Reconciled? How Biological Anthropologists View Human Variation'. *American Journal of Physical Anthropology* 139: 1–4. https://doi.org/10.1002/ajpa.20995.

Elkin, A.P. 1974. 'Elliot Smith and the Diffusion of Culture'. In *Grafton Elliot Smith: The Man and His Work*, edited by A.P. Elkin and N.W.G. Macintosh, 139–159. Sydney: Sydney University Press.

Ellison, George T.H., Thea de Wet, Carel B. IJsselmuiden and Linda M. Richter. 1996. 'Desegregating Health Statistics and Health Research in South Africa'. *South African Medical Journal* 86, no. 10: 1257–1262.

Elsdon-Dew, Ronald. 1935. 'The Blood Groups of the Hottentot'. *South African Medical Journal* 9, no. 18: 651–653. Accessed 30 July 2021. https://hdl.handle.net/10520/AJA20785135_11727.

Erasmus, Yvonne. 2007. 'Racial (Re)Classification during Apartheid South Africa: Regulations, Experiences and the Meaning(s) of "Race"'. PhD diss. University of London.

Erasmus, Yvonne and George T.H. Ellison. 2008. 'What Can We Learn about the Meaning of Race from the Classification of Population Groups during Apartheid?' *South African Journal of Science* 104: 450–452. http://dx.doi.org/10.1590/S0038-23532008000600016.

Eriksson, Gunnar. 1983. 'Linnaeus the Botanist'. In *Linnaeus: The Man and His Work*, edited by Tore Frängsmyr, 63–109. Berkeley: University of California Press.

Farini, Gilarmi A. 1973. *Through the Kalahari Desert*. Cape Town: Struik. Reprint of 1886 edition.

Findlay, George. 1972. *Dr. Robert Broom: Palaeontologist and Physician, 1866–1951*. Cape Town: Balkema.

Fischer, Eugen. 1913. *Die Rehobother Bastards und das Bastardierungsproblem beim Menschen*. Jena: Gustav Fischer.

FitzSimons, Frederick W. 1911. *The Monkey Folk of South Africa*. London: Longmans Green.

FitzSimons, Frederick W. 1921. 'The Cliff-Dwellers of Tzitzikama'. *Illustrated London News*, 24 December: 880.

FitzSimons, Frederick W. 1923. 'Cliff Dwellers of Zitzikama'. *South African Journal of Science* 20: 541–544.

FitzSimons, Frederick W. 1926. 'Cliff Dwellers of Zitzikama: Results of Recent Excavations'. *South African Journal of Science* 23: 813–817. Accessed 30 July 2021. https://hdl.handle.net/10520/AJA00382353_10130.

Fouché, Leo. 1937. *Mapungubwe: Ancient Bantu Civilization on the Limpopo*. Cambridge: Cambridge University Press.

Galloway, Alexander. 1933. 'The Nebarara Skull'. *South African Journal of Science* 30: 585–596. https://journals.co.za/doi/pdf/10.10520/AJA00382353_4365.

Galloway, Alexander. 1934. 'A Note on the Iron Smelting Methods of the Elgeyo Masai'. *South African Journal of Science* 31: 500–504. Accessed 30 July 2021. https://hdl.handle.net/10520/AJA00382353_6458.

Galloway, Alexander. 1935. 'Stone Structures on the Uasin Gishu Plateau, Kenya Colony'. *South African Journal of Science* 32: 656–658.

Galloway, Alexander. 1937a. 'The Characteristics of the Skull of the Boskop Physical Type'. *American Journal of Physical Anthropology* 23, no. 1: 31–46. https://doi.org/10.1002/ajpa.1330230105.

Galloway, Alexander. 1937b. 'A Contribution to the Physical Anthropology of the Ovambo'. *South African Journal of Science* 34: 351–364. https://journals.co.za/doi/pdf/10.10520/AJA00382353_8831.

Galloway, Alexander. 1937c. 'The Skeletal Remains of Mapungubwe'. In *Mapungubwe*, edited by Leo Fouché, 127–174. Cambridge: Cambridge University Press.

Galloway, Alexander. 1959. *The Skeletal Remains of Bambandyanalo*. Johannesburg: Wits University Press.

Gates, Reginald Ruggles. 1926. 'Mendelian Heredity and Racial Differences'. *Journal of the Royal Anthropological Institute* 55: 468–482. https://doi.org/10.2307/2843651.

Gates, Reginald Ruggles. 1929. *Heredity in Man*. London: Constable & Co.

Gates, Reginald Ruggles. 1948. *Human Ancestry: From a Genetical Point of View*. Cambridge, MA: Harvard University Press.

Gear, Harry Sutherland. 1925. 'The Skeletal Features of the Boskop Race'. *South African Journal of Science* 22: 458–469. Accessed 30 July 2021. https://hdl.handle.net/10520/AJA00382353_8596.

Gear, Harry Sutherland. 1926. 'A Further Report on the Boskopoid Remains from Zitzikama'. *South African Journal of Science* 23: 923–934. Accessed 30 July 2021. https://hdl.handle.net/10520/AJA00382353_10144.

Gear, John Hallward. 1929. 'Cranial Form in the Native Races of South Africa'. *South African Journal of Science* 26: 684–697. Accessed 30 July 2021. https://hdl.handle.net/10520/AJA00382353_4037.

Girvin, Stephen D. 1987. 'Race and Race Classification'. In *Race and the Law in South Africa*, edited by A. Rycroft, 1–10. Cape Town: Juta.

Glaser, S. 1934. 'Sweat Glands in the Negro and the European'. *American Journal of Physical Anthropology* 18, no. 3: 371–376. https://doi.org/10.1002/ajpa.1330180318.

Goldby, Frank. 1956. 'In Memoriam: Dr Wynfrid Lawrence Henry Duckworth'. *Journal of Anatomy* 90, no. 3: 455–456.

Goodwin, A. John H. 1935. 'A Commentary on the History and Present Position of South African Prehistory with Full Bibliography'. *Bantu Studies* 9, no. 1: 291–417. https://doi.org/10.1080/02561751.1935.9676376.

Goodwin, A. John H. and Clarence van Riet Lowe. 1929. 'The Stone Age Cultures of South Africa'. *Annals of the South African Museum* 27: 1–289.

Gordon, Robert. 1988. 'Apartheid's Anthropologists: The Genealogy of Afrikaner Anthropology'. *American Ethnologist* 15, no. 3: 335–353. https://doi.org/10.1525/ae.1988.15.3.02a00090.

Gould, Stephen Jay. 1978. 'Morton's Ranking of Races by Cranial Capacity'. *Science* 200, no. 4341: 503–509. https://doi.org/10.1126/science.347573.

Gowlett, John A.J. 1990. 'Archaeological Studies of Human Origins and Early Prehistory in Africa'. In A *History of African Archaeology*, edited by Peter Robershaw, 13–38. London: James Currey.

Grintzesco, Suzanne. 1933. 'Contribution à l'étude anthropologique du femur des Boschimans'. PhD diss., University of Geneva.

Grobbelaar, Coert. 1955. 'The Distribution of the Blood Groups of the Koranas'. *South African Journal of Science* 53: 323–326. Accessed 30 July 2021. https://hdl.handle.net/10520/AJA00382353_6648.

Guenther, Lars and Peter Weingart. 2016. 'A Unique Fingerprint? Factors Influencing Attitudes towards Science and Technology in South Africa'. *South African Journal of Science* 112, no. 7–8. https://doi.org/10.17159/sajs.2016/20160093.

Haddon, Alfred Cort. 1909. *The Races of Man and Their Distribution*. London: Milner & Co.

Haffajee, Ferial. 2020. 'The Stubborn Stickiness of Race Science in SA in the 21st Century'. *Daily Maverick*, 29 June.

Hall, Martin. 1990. 'Hidden History: Iron Age Archaeology in Southern Africa'. In A *History of African Archaeology*, edited by Peter Robershaw, 59–77. London: James Currey.

Hall, Martin and Alan G. Morris. 1983. 'Race and Iron Age Human Skeletal Remains from Southern Africa: An Assessment'. *Social Dynamics* 9, no. 2: 29–36. https://doi.org/10.1080/02533958308458344.

Hamer, William Heaton. 1935. 'Obituary: F.C. Shrubsall'. *British Medical Journal*, 5 October: 646–647.

Hartshorne, Kenneth. 1992. *Crisis and Challenge: Black Education, 1910–1990*. Cape Town: Oxford University Press.

Haughton, Sidney H., Robert B. Thomson and Louis Péringuey. 1917. 'Preliminary Note on the Ancient Human-Skull Remains from the Transvaal'. *Transactions of the Royal Society of South Africa* 6, no. 1: 1–13. https://doi.org/10.1080/00359191709520168.

Hawks, John. 2008. 'The "Amazing" Boskops'. *john hawks weblog* (blog), 30 March. Accessed 19 July 2021. http://johnhawks.net/weblog/reviews/brain/paleo/lynch-granger-big-brain-boskops-2008.html.

Hawks, John. 2010. '"The Return of the "Amazing" Boskops'. *john hawks weblog* (blog), 4 January. Accessed 19 July 2021. http://johnhawks.net/weblog/reviews/brain/paleo/return-amazing-boskops-lynch-granger-2009.html.

Henderson, Zoë. 1992. 'The Florisbad Skull: A Diamond Jubilee'. *Culna* 43: 18–19.

Henderson, Zoë. 1994. 'Captain R. Egerton Helme: A Benefactor of Research'. *Culna* 46: 28–29.

Herrnstein, Richard J. and Charles Murray. 1994. *The Bell Curve: Intelligence and Class Structure in American Life*. New York: Free Press.

Hewitt, John. 1921. 'On Several Implements and Ornaments from Strandloper Sites in the Eastern Province'. *South African Journal of Science* 18: 454–467.

Hiernaux, Jean. 1969. 'Biological Aspects of the Racial Question'. In *Four Statements on the Race Question*, 9–16. Paris: Unesco. Accessed 19 July 2021. http://refugeestudies.org/UNHCR/UNHCR.%20Four%20Statements%20on%20the%20Race%20Question.pdf.

Hildebrandt, Sabine. 2009. 'Anatomy in the Third Reich: An Outline, Part 2 – Bodies for Anatomy and Related Medical Disciplines'. *Clinical Anatomy* 22, no. 8: 894–905. https://doi.org/10.1002/ca.20873.

Hoffman, Abraham C. 1955a. 'Important Contributions of the Orange Free State to Our Knowledge of Primitive Man'. *South African Journal of Science* 51, no. 6: 163–168. Accessed 30 July 2021. https://hdl.handle.net/10520/AJA00382353_449.

Hoffman, Abraham C. 1955b. 'Obituary: T.F. Dreyer'. *Navorsinge van die Nasionale Museum* 1, no. 8: 185–186. Accessed 30 July 2021. https://hdl.handle.net/10520/AJA00679208_142.

Hoffman, Abraham C. 1960. 'The Hoffman Angola Expedition'. *South African Journal of Science* 56, no. 3: 65–66. Accessed 30 July 2021. https://hdl.handle.net/10520/AJA00382353_8632.

Hoffman, Abraham C. 1972. 'Kakamas Burial Sites'. *Standard Encyclopedia of Southern Africa* 6: 274–275.

Horn, David B. 1967. *A Short History of the University of Edinburgh*. Edinburgh: Edinburgh University Press.

Howells, William W. 1973. *Cranial Variation in Man*. Papers of the Peabody Museum of Archaeology and Ethnology, volume 67. Cambridge, MA: Harvard University Press.

Howells, William W. 1989. *Skull Shape and the Map: Craniometric Analyses in the Dispersion of Modern* Homo. Cambridge, MA: Harvard University Press.

Horrell, Muriel. 1953. *A Survey of Race Relations in South Africa, 1952–1953*. Johannesburg: South African Institute of Race Relations.

Horrell, Muriel. 1957. *A Survey of Race Relations in South Africa, 1956–1957*. Johannesburg: South African Institute of Race Relations.

Horrell, Muriel. 1966. *A Survey of Race Relations in South Africa, 1965–1966*. Johannesburg: South African Institute of Race Relations.

Horrell, Muriel. 1967. *A Survey of Race Relations in South Africa, 1966–1967*. Johannesburg: South African Institute of Race Relations.

Horrell, Muriel. 1972. *A Survey of Race Relations in South Africa, 1971–1972*. Johannesburg: South African Institute of Race Relations.

Horrell, Muriel. 1973. *A Survey of Race Relations in South Africa, 1972–1973*. Johannesburg: South African Institute of Race Relations.

Horrell, Muriel. 1974. *A Survey of Race Relations in South Africa, 1973–1974*. Johannesburg: South African Institute of Race Relations.

Horrell, Muriel. 1975. *A Survey of Race Relations in South Africa, 1974–1975*. Johannesburg: South African Institute of Race Relations.

Horrell, Muriel. 1976. *A Survey of Race Relations in South Africa, 1975–1976*. Johannesburg: South African Institute of Race Relations.

Horrell, Muriel. 1977. *A Survey of Race Relations in South Africa, 1976–1977*. Johannesburg: South African Institute of Race Relations.

Horrell, Muriel. 1978. *A Survey of Race Relations in South Africa, 1977–1978*. Johannesburg: South African Institute of Race Relations.

Horrell, Muriel. 1979. *A Survey of Race Relations in South Africa, 1978–1979*. Johannesburg: South African Institute of Race Relations.

Hull, David L. 1988. *Science as a Process*. Chicago: University of Chicago Press.

Hultkrantz, Ake. 1980. 'Anthropological Traditions: Comparative Aspects'. In *Anthropology: Ancestors and Heirs*, edited by Stanley Diamond, 89–105. The Hague: Mouton.

Humphreys, Anthony J.B. 1972. 'The Type R Settlements in the Context of the Later Prehistory and Early History of the Riet River Valley'. Master's thesis, University of Cape Town.

Hunley, Keith L., Meghan E. Healy and Jeffrey C. Long. 2009. 'The Global Pattern of Gene Identity Variation Reveals a History of Long-Range Migrations, Bottlenecks, and Local Mate Exchange: Implications for Biological Race'. *American Journal of Physical Anthropology* 139, no. 1: 35–46. https://doi. org/10.1002/ajpa.20932.

Huxley, Julian. 1942. *Evolution: The Modern Synthesis*. London: George Allen & Unwin.

Huxley, Julian, Alfred C. Haddon and Alexander M. Carr-Saunders. 1935. *We Europeans: A Survey of 'Racial' Problems*. London: Penguin.

Jacobs, Zenobia, Richard G. Roberts, Rex F. Galbraith, Hilary J. Deacon, Rainer Grün, Alex Mackay, Peter Mitchell, Ralf Vogelsang and Lyn Wadley. 2008. 'Ages for the Middle Stone Age of Southern Africa: Implications for Human Behavior and Dispersal'. *Science* 322, no. 5902: 733–735. https://doi.org/10.1126/ science.1162219.

Jenkins, Trefor and Phillip V. Tobias. 1977. 'Nomenclature of Population Groups in Southern Africa'. *African Studies* 36, no. 1: 49–55. https://doi.org/10.1080/ 00020187708707485.

Jones, Kerry. 2003. 'Contemporary Khoesan Languages of South Africa'. *Critical Arts* 33, no. 4–5: 33–54. https://doi.org/10.1080/02560046.2019.1688849.

Kaufman, Pamela de Beer. 1975. 'A Study of the Sacrum and Some Aspects of Presacral Vertebrae in San (Bushmen) and Southern African and American Negroes'. PhD diss., University of the Witwatersrand.

Keen, Edward N. 1949. 'A Method of Measuring the Profile Angle at the Nasion in Skulls'. *South African Journal of Medical Science* 14: 7–12.

Keen, Edward N. 1953. 'Estimation of Stature from the Long Bones'. *Journal of Forensic Medicine* 1: 46–51.

Keen, Jack A. 1941. 'Report on a Skeleton from the Fish Hoek Cave'. *South African Journal of Science* 38: 301–309. Accessed 30 July 2021. https://hdl.handle. net/10520/AJA00382353_8337.

Keen, Jack A. 1950. 'A Study of the Differences between Male and Female Skulls'. *American Journal of Physical Anthropology* 8, no. 1: 65–80. https://doi.org/ 10.1002/ajpa.1330080113.

Keen, Jack A. 1951. 'Craniometric Study of the Cape Coloured Population'. *Transactions of the Royal Society of South Africa* 33, no. 1: 29–51. https://doi.org/ 10.1080/00359195109519876.

Keene, Rochelle. 2013. *Our Graduates, 1925–2012*. Johannesburg: Adler Museum of Medical History.

Keith, Arthur. 1925. *The Antiquity of Man*. 2nd edition. London: Williams & Norgate.

Keith, Arthur. 1931. *New Discoveries Relating to the Antiquity of Man*. New York: Norton.

Kirby, Jeremy. 2009. 'Review of *Big Brain*, by G. Lynch and R. Granger'. *Quarterly Review of Biology* 84, no. 2: 196. https://doi.org/10.1086/603481.

Kuljian, Christa. 2016. *Darwin's Hunch: Science, Race and the Search for Human Origins*. Johannesburg: Jacana Media.

Kuper, Adam. 1983. *Anthropology and Anthropologists*. 2nd edition. London: Routledge & Kegan Paul.

L'Abbé, Ericka N., Marius Loots and Natalie Keough. 2008. 'The Matjes River Rock Shelter: A Description of the Skeletal Assemblage'. *South African Archaeological Bulletin* 63, no. 187: 61–68. https://doi.org/10.2307/20474993.

Laing, Gordon D. 1924. 'A Preliminary Report on Some Strandlooper Skulls Found at Zitzikama'. *South African Journal of Science* 21: 528–541. Accessed 30 July 2021. https://hdl.handle.net/10520/AJA00382353_2576.

Laing, Gordon D. 1925. 'A Further Report on the Zitzikama Material'. *South African Journal of Science* 22: 455–457. Accessed 30 July 2021. https://hdl.handle. net/10520/AJA00382353_8595.

Laing, Gordon D. 1926. 'The Relationship between Boskop, Bushman and Negro Elements in the Formation of the Native Races of South Africa'. *South African Journal of Science* 23: 905–908. Accessed 30 July 2021. https://hdl.handle. net/10520/AJA00382353_10142.

Laing, Gordon D. and Harry S. Gear. 1929. 'A Final Report on the Strandlooper Skulls Found at Zitzikama'. *South African Journal of Science* 26: 575–601. Accessed 30 July 2021. https://hdl.handle.net/10520/AJA00382353_4030.

Le Grange, Lesley. 2019. 'A Comment on Critiques of the Article "Age- and Education-Related Effects on Cognitive Functioning in Colored South African Women"'. *South African Journal of Higher Education* 33, no. 4: 9–19. https://doi. org/10.20853/33-4-3715.

Legassick, Martin and Cirraj Rassool. 2000. *Skeletons in the Cupboard: Museums and the Incipient Trade in Human Remains, 1907–1917*. Cape Town: South African Museum.

Lewontin, Richard C. 1972. 'The Apportionment of Human Diversity'. In *Evolutionary Biology 6*, edited by Theodosius Dobzhansky, Max K. Hecht and William C. Steere, 381–398. New York: Springer. https://doi.org/10.1007/978-1-4684-9063-3_14.

Lindroth, Sten. 1983. 'The Two Faces of Linnaeus'. In *Linnaeus: The Man and His Work*, edited by Tore Frängsmyr, 1–62. Berkeley: University of California Press.

Little, Michael A. and Kenneth J. Collins. 2012. 'Joseph S. Weiner and the Foundation of Post-WWII Human Biology in the United Kingdom'. *Yearbook of Physical Anthropology* 149, no. S55: 114–131. https://doi.org/10.1002/ajpa.22164.

Littlefield, Alice, Leonard Lieberman and Larry T. Reynolds. 1982. 'Redefining Race: The Potential Demise of a Concept in Physical Anthropology'. *Current Anthropology* 23, no. 6: 641–655. https://doi.org/10.1086/202915.

Lounsbury, C.P. 1940. 'The Pioneer Period of Economic Entomology in South Africa'. *Journal of the Entomological Society of Southern Africa* 3: 9–29. Accessed 19 July 2021. https://journals.co.za/doi/pdf/10.10520/AJA00128789_2400.

Louw, Jan H. 1969. *In the Shadow of Table Mountain: A History of the University of Cape Town Medical School.* Cape Town: Struik.

Louw, J.T. 1960. *Prehistory of the Matjes River Rock Shelter.* National Museum, Memoir No. 1. Bloemfontein: National Museum.

Lynch, Gary and Richard Granger. 2008. *Big Brain: The Origins and Future of Human Intelligence.* New York: St Martin's Press.

Lynch, Gary and Richard Granger. 2009. 'What Happened to Hominids Who Were Smarter Than Us?' *Discover*, 28 December. Accessed 19 July 2021. https://www.discovermagazine.com/mind/what-happened-to-the-hominids-who-may-have-been-smarter-than-us.

MacDonald, Helen. 2010. *Possessing the Dead: The Artful Science of Anatomy.* Melbourne: Melbourne University Press.

Magner, Lois N. 1979. *A History of the Life Sciences.* New York: Dekker.

Maingard, J.F. 1932. 'Physical Characteristics of the Korana'. *Bantu Studies* 6, no. 1: 163–182. https://doi.org/10.1080/02561751.1932.9676281.

Maingard, Louis F. 1932. 'Studies in Korana History, Customs and Language'. *Bantu Studies* 6, no. 1: 103–121. https://doi.org/10.1080/02561751.1932.9676280.

Marks, Jonathan. 2009. *Why I Am Not a Scientist: Anthropology and Modern Knowledge.* Berkeley: University of California Press.

Martinaglia, André. 2013. 'In Africa from the Cape to Cairo with Lidio Cipriani'. *La Gazzetta del Sud Africa Quotidiano indipendente d'informazione degli Italiani del Sud Africa.* Accessed 22 March 2016. http://www.lagazzettadelsudafrica.net/lidio-cipriani.html.

Martinez, Susan B. 2016. *The Lost Continent of Pan.* New York: Simon & Schuster.

Mason, Revil J. 1989. 'South African Archaeology, 1922–1988'. Occasional Paper No. 22, Archaeological Research Unit, University of the Witwatersrand.

Mathiane, Nomavenda. 1990. *Beyond the Headlines: Truths of Soweto Life.* Johannesburg: Southern Book Publishers.

Matus, S. 1941. 'The Mongol Spot in the Cape Coloured'. *South African Medical Journal* 12 April: 15. Accessed 30 July 2021. https://hdl.handle.net/10520/AJA20785135_17689.

Mazel, Aron. 2013. 'Apartheid's Child: The Creation of the South African Cultural History Museum in the 1950s and 1960s'. *Museum History Journal* 6, no. 2: 166–202. https://doi.org/10.1179/1936981613Z.00000000012.

McKittrick, Meredith. 2015. 'An Empire of Rivers: The Scheme to Flood the Kalahari, 1919–1945'. *Journal of Southern African Studies* 41, no. 3: 485–504. https://doi.org/10.1080/03057070.2015.1025339.

Meier, Robert J. 2018. 'J.S. Weiner'. In *International Encyclopedia of Biological Anthropology*, edited by Wenda Trevathan. New York: John Wiley & Sons. http://dx.doi.org/10.1002/9781118584538.ieba0518.

Meiring, Albert J.D. 1937. 'The "Wilton" Skulls of the Matjes River Shelter'. *Soologiese Navorsing van die Nasionale Museum* 1, no. 6: 51–79.

Midgley, Jeremy J. 2020. 'A Question Worth Asking'. *South African Journal of Science* 116 (special issue): 22–24. https://doi.org/10.17159/sajs.2020/8591.

Mnguni, Lindelani. 2020. '#BlackLivesMatter, Even in Research: A Call to Researchers to Take a Knee'. *South African Journal of Science* 116 (special issue): 25–29. https://doi.org/10.17159/sajs.2020/8540.

Montagu, Ashley. 1942. *Man's Most Dangerous Myth: The Fallacy of Race*. New York: Columbia University Press.

Montagu, Ashley. 1951. *Statement on Race*. New York: Henry Schuman.

Morris, Alan G. 1980. 'The Role of Physical Anthropology in the Study of African History'. Paper presented at the Centre for African Studies Seminar, University of Cape Town, 15 October.

Morris, Alan G. 1984. 'An Osteological Analysis of the Protohistoric Populations of the Northern Cape and Western Orange Free State, South Africa'. PhD diss., University of the Witwatersrand.

Morris, Alan G. 1986. 'Khoi and San Craniology: A Re-evaluation of the Osteological Reference Samples'. In *Variation, Culture and Evolution in African Populations: Papers in Honour of Dr Hertha de Villiers*, edited by Ronald Singer and John K. Lundy, 1–12. Johannesburg: Wits University Press.

Morris, Alan G. 1987. 'The Reflection of the Collector: San and Khoi Skeletons in Museum Collections'. *South African Archaeological Bulletin* 42, no. 145: 12–22. https://doi.org/10.2307/3887769.

Morris, Alan G. 1988a. 'Discussing Race in a Racist Society'. *Anthropology Today* 4, no. 1: 3–5. https://doi.org/10.2307/3032870. Reprinted in *World Archaeological Bulletin* 3 (1989), with a rejoinder from G.T. Nurse.

Morris, Alan G. 1988b. 'Discussing Race in a Race-Conscious Society: Scientific and Lay Perceptions of Human Variation in South Africa'. Paper presented at the Royal Society of South Africa Seminar, Cape Town, 16 March.

Morris, Alan G. 1992a. *A Master Catalogue: Holocene Human Skeletons from South Africa*. Johannesburg: Wits University Press.

Morris, Alan G. 1992b. *The Skeletons of Contact: Protohistoric Burials from the Lower Orange River Valley*. Johannesburg: Wits University Press.

Morris, Alan G. 1995. 'The Einiqua: An Analysis of the Kakamas Skeletons'. In *Einiqualand: Studies of the Orange River Frontier*, edited by Andrew B. Smith, 110–164. Cape Town: University of Cape Town Press.

Morris, Alan G. 2000. 'Comment'. In *Skeletons in the Cupboard: Museums and the Incipient Trade in Human Remains, 1907–1917*, by Martin Legassick and Ciraj Rassool, 73–76. Cape Town: South African Museum.

Morris, Alan G. 2002. 'The British Association Meeting of 1905 and the Rise of Physical Anthropology in South Africa'. *South African Journal of Science* 98, no. 7: 336–340. Accessed 19 July 2021. https://journals.co.za/doi/pdf/10.10520/EJC97516.

Morris, Alan G. 2003. 'Using Racial Terms in Anatomical Research'. *Plexus: Newsletter of the International Federation of Associations of Anatomists* 2 (May).

Morris, Alan G. 2004. 'Anatomy and Archaeology: A Report on the World Archaeo-logical Congress 5, Washington, DC, 21 to 26 June, 2003'. *Plexus: Newsletter of the International Federation of Associations of Anatomists* 3 (April): 3–5.

Morris, Alan G. 2005a. 'Dead White Men: Scientific Racism in Context'. In *Yes, Dead Men Do Tell Tales!*, edited by Louise J. Friedling, 7–20. Cape Town: David and Elaine Potter Charitable Trust and the South Africa-Netherlands Research Pro-gramme on Alternatives in Development (SANPAD).

Morris, Alan G. 2005b. 'Measure by Measure: The History of Race and Typology in South African Physical Anthropology'. In *Voyages in Science: Essays by South African Anatomists in Honour of Phillip V. Tobias' 80th Birthday*, edited by Goran Štrkalj, N. Pather and Beverley Kramer, 121–140. Pretoria: Content Solutions.

Morris, Alan G. 2005c. 'Prehistory in Blood and Bone: An Essay on the Reconstruc-tion of the Past from Genetics and Morphology'. *Transactions of the Royal Society of South Africa* 60, no. 2: 111–114. https://doi.org/10.1080/00359190509520487.

Morris, Alan G. 2009a. 'Burning the Straw Man: A Response to Hall's "New Know-ledge and the University"'. *Anthropology Southern Africa* 32, no. 1–2: 78–79.

Morris, Alan G. 2009b. 'Zuckerman versus Marais: A Primatological Collision'. *South African Journal of Science* 105, no. 5–6: 238–240. Accessed 19 July 2021. http://www.scielo.org.za/pdf/sajs/v105n5-6/a1810556.pdf.

Morris, Alan G. 2010a. 'From Snakes to Skeletons: A Personal Voyage of Scientific Discovery'. Unpublished manuscript.

Morris, Alan G. 2010b. 'Racial Identification of Single Skulls in Forensic Cases: When Myth Becomes Reality'. In *Teaching Human Variation: Issues, Trends and Challenges*, edited by Goran Štrkalj, 15–31. Hauppauge, NY: Nova Science Publishing.

Morris, Alan G. 2011. *Missing and Murdered: A Personal Adventure in Forensic Anthropology*. Cape Town: Zebra Press.

Morris, Alan G. 2012. 'Biological Anthropology at the Southern Tip of Africa: Carrying European Baggage in an African Context'. *Current Anthropology* 53, no. S5: S152–S160. https://doi.org/10.1086/662289.

Morris, Alan G. and Edward N. Keen. 1982. 'Lawrence Herbert Wells and the His-tory of Anatomical Illustration'. *South African Medical Journal* 61, no. 2: 40–43. Accessed 30 July 2021. https://hdl.handle.net/10520/AJA20785135_14299.

Morris, Alan G., Vince Phillips and Susan Rosendorff. 2012. 'Facial Reconstruction at Cobern Street'. Paper presented at Anatomical Association of Southern Africa meeting, Windhoek, 15 April.

Morris, Alan G. and Phillip V. Tobias. 1997. 'South Africa'. In *The Encyclopaedia of the History of Physical Anthropology*, edited by Frank Spencer, 968–976. New York: Garland.

Mosothwane, Morongwa N. 2013. 'An Account of South African Human Skeletal Remains at Three North American Museum Collections'. *South African Arch-aeological Society, Goodwin Series* 11: 27–34. Accessed 19 July 2021. https://www.jstor.org/stable/43997026.

Mossop, E.E. (ed.). 1935. *The Journal of Hendrik Jacob Wikar (1779) with an English Translation by A.W. van der Horst and the Journals of Jacobus Coetsé Jansz (1760) and Willem van Reenen (1791) with an English Translation by Dr. E.E. Mossop*. Cape Town: Van Riebeeck Society.

Murray, Bruce K. 1982. *Wits, the Early Years*. Johannesburg: Wits University Press.

Murray, Bruce K. 1997. *Wits: The 'Open' Years*. Johannesburg: Wits University Press.

Nattrass, Nicoli. 2020. 'Why Are Black South African Students Less Likely to Consider Studying Biological Sciences?' *South African Journal of Science* 116, no. 5–6: https://doi.org/10.17159/sajs.2020/7864.

Ncayiyana, Daniel J. 2007. 'Racial Profiling in Medical Research: What Are We Measuring?' *South African Medical Journal* 97, no. 12: 1225–1226. Accessed 19 July 2021. https://hdl.handle.net/10520/EJC68943.

Nieuwoudt, Sharné, Kasha E. Dickie, Carla Coetsee, Louise Engelbrecht and Elmarie Terblanche. 2020. RETRACTED ARTICLE. 'Age- and Education-Related Effects on Cognitive Functioning in Colored South African Women'. *Aging, Neuropsychology, and Cognition* 27, no. 3: 321–337. https://doi.org/10.1080/13 825585.2019.1598538. ['Statement of Retraction: Age- and Education-Related Effects on Cognitive Functioning in Colored South African Women'. *Aging, Neuropsychology, and Cognition* 27, no. 6 (2020): 963. https://doi.org/10.1080/1 3825585.2019.1614759.]

Nurse, George T. 1975. 'Origins of the Northern Cape Griqua'. ISMA Paper No. 34. Johannesburg: Institute for the Study of Man. Accessed 19 July 2021. http://www.historicalpapers.wits.ac.za/inventories/inv_pdfo/A1132/A1132-C407-001-jpeg.pdf.

Nurse, George T. 1976. 'Isonymic Studies on the Griqua of the Northern Cape Province of South Africa'. *Journal of Biosocial Science* 8, no. 3: 277–286. https://doi.org/10.1017/s0021932000010750.

Nurse, George T. 1988. 'Letter to the Editor'. *Anthropology Today* 4, no. 3: 24–25.

Nurse, George T. and Trefor Jenkins. 1975. 'The Griqua of Campbell, Cape Province, South Africa'. *American Journal of Physical Anthropology* 43: 71–78. https://doi.org/10.1002/ajpa.1330430111.

Nurse, George T., Joseph Weiner and Trefor Jenkins 1985. *The Peoples of Southern Africa and Their Affinities*. Oxford: Oxford University Press.

Obituary. 1965. 'Alexander Galloway'. *The Lancet* 2 (9 October): 748–749.

Obituary. 1966. 'Professor Matthew Robertson Drennan'. *South African Journal of Science* 62, no. 4: 122–125.

Obituary. 1986. 'Professor Meiring'. *Die Bult*: 2.

Orford, Margaret and Lawrence H. Wells. 1937. 'An Anthropometric Study of a Series of South African Bantu Females'. *South African Journal of Science* 33: 1010–1036. Accessed 30 July 2021. https://hdl.handle.net/10520/AJA00382353_8213.

Orkin, Carole. 1970. 'The Tablier in Griqua Women'. *The Leech* 40: 42–43.

Pankhurst, Richard. 2005. 'Racism in the Service of Fascism, Empire-Building and War: The History of the Italian Fascist Magazine "La Difesa della Razza"'. In *Auf dem Weg zum modernen Äthiopien, Festschrift für Bairu Tafla*, edited by Stefan Brüne and Heinrich Scholler, 134–163. Münster: Lit Verlag. Accessed 19 July 2021. https://www.marxists.org/archive/pankhurst-richard/2007/03/x01.htm.

Pearson, Karl. 1925. 'Foreword'. *Annals of Eugenics* 1, no. 1: 1–4. https://doi.org/10.1111/j.1469-1809.1925.tb02036.x.

Pearson, Karl. 1926. 'On the Coefficient of Racial Likeness'. *Biometrika* 18, no. 1–2: 105–117. https://doi.org/10.2307/2332498.

Peckmann, Tanya R. 2002. 'Dialogues with the Dead: An Osteological Analysis of the Palaeodemography and Life History of the 18th and 19th Century Northern Frontier in South Africa'. PhD diss., University of Cape Town.

Penniman, T. Kenneth. 1952. *A Hundred Years of Anthropology*. 2nd edition. London: Duckworth & Co.

Penrose, Lionel S. 1952. 'Distance, Size and Shape'. *Annals of Eugenics* 17, no. 1: 337–343. https://doi.org/10.1111/j.1469-1809.1952.tb02527.x.

Perez, A.M., N. Ahmed and Leslie London. 2012. 'Racial Discrimination: Experiences of Black Medical School Alumni at the University of Cape Town, 1945–1994'. *South African Medical Journal* 102, no. 6: 574–577. https://doi.org/10.7196/samj.5721.

Périer, A. 1930. 'A propos du tubercle de Carabelli, sa présonce chez les Boschimans – Hottentots – Griquas et quelques remarques sur sa signification'. *Archives Suisses d'Anthropologie Generale* 5, no. 2: 174–178.

Péringuey Louis. 1905. 'The Stone Age in South Africa'. In *Science in South Africa: A Handbook and Review*, edited by W. Flint and J.D.F. Gilchrist, 102–108. Cape Town: Maskew Miller.

Péringuey, Louis. 1911. 'The Stone Ages of South Africa as Represented in the Collection of the South African Museum'. *Annals of the South African Museum* 8: 1–174.

Péringuey, Louis and George S. Corstorphine. 1900. 'Stone Implements from Bosman's Crossing, Stellenbosch'. In *Minutes of the Proceedings of the South African Philosophical Society* 11: xxiv.

Petersen, Arne F. 2004. 'Niels Bolwig: Behavioural Biologist Who Was One of the First to Study Primates in the Wild'. *The Independent*, 17 November.

Phillips, Howard. 2019. *UCT under Apartheid: Part 1, 1948–1968*. Johannesburg: Jacana Media.

Piccioni, Lucia. 2020. 'The Facial Casts of the Anthropology Museum in Florence'. *Politika*, 8 January. Accessed 19 July 2021. https://www.politika.io/en/notice/duplicating-and-hierarchizing-humanity.

Pijper, Adrianus. 1930. 'The Blood Groups of the Bantu'. *Transactions of the Royal Society of South Africa* 18, no. 4: 311–315. https://doi.org/10.1080/00359193009518812.

Pittard, Eugène. 1910. *Crania Helvetica: Les cranes Valaisans de la Vallée du Rhone*. Geneva: Mémoires de l'Institut national genevois.

Pittard, Eugène. 1916. *Les race belligérantes: Esquisses anthropologiques*. Paris: La Renaissance du Livre.

Pittard, Eugène. 1924. *Les races et l'histoire: Introduction ethnologique à l'histoire*. Paris: La Renaissance du Livre.

Pittard, Eugène. 1925/26. 'L'indice céphalique des Boschimans'. *Volume jubilaire de Gorjanovic-Kramberger*: 533–551.

Pittard, Eugène. 1939. 'Quelques observations au sujet des trous pariétaux chez les crâne des Boschimans, des Hottentots et des Griquas'. *Archives Suisses d'Anthropologie Generale* 8, no. 2: 172–186.

Pittard, Eugène and M. Baïcoyano. 1928. 'L'indice dentaire de Flower chez les Boschimens, Hottentots et Griquas. Comparison sexuelles'. In *Institut International d'Anthropologie*, 3rd session. Amsterdam, 20–29 September 1927: 260–264.

Pittard, Eugène and Juan Comas. 1930. 'La reconstitution de la taille à l'aide du femur dans une série de squelettes de Boschimans, Hottentots et Griquas'. *Archives Suisses d'Anthropologie Generale* 5, no. 2: 248–250.

Plug, Cornelis. 2020a. 'Dreyer, Prof. Thomas Frederik'. *S2A3 Biographical Database of Southern African Science*. Accessed 24 April 2020. https://www.s2a3.org.za/bio/Biograph_final.php?serial=779.

Plug, Cornelis. 2020b. 'Frederick William FitzSimons'. *S2A3 Biographical Database of Southern African Science*. Accessed 24 April 2020. http://www.s2a3.org.za/bio/Biograph_final.php?serial=937.

Plug, Cornelis. 2020c. 'Mr James Drury'. *S2A3 Biographical Database of Southern African Science*. Accessed 16 May 2020. http://www.s2a3.org.za/bio/Biograph_final.php?serial=784.

Plug, Cornelis. 2020d. 'Péringuey, Dr. Louis Albert'. *S2A3 Biographical Database of Southern African Science*. Accessed 24 April 2020. https://www.s2a3.org.za/bio/Biograph_final.php?serial=2170.

Pöch, Rudolf. 1909. 'Untersuchung Von Buschmann: Schadeln und Skeletten aus dem Transvaal Museum'. *Annals of the Transvaal Museum* 1: 199–218.

Popescu-Spineni, Dana. 2015. 'An Outstanding Personality of European Culture in Romania: Eugène Pittard'. *Proceedings of the Romanian Academy* 17, no. 2: 157–164. Accessed 19 July 2021. https://acad.ro/sectii2002/proceedingsChemistry/doc2015-2/art07Spineni.pdf.

Posel, Deborah. 2001. 'Race as Common Sense: Racial Classification in Twentieth-Century South Africa'. *African Studies Review* 44, no. 2: 87–113. https://doi.org/10.2307/525576.

Proctor, Robert. 1988. 'From Anthropologie to Rassenkunde in the German Anthropological Tradition'. In *Bones, Bodies, Behavior: Essays on Biological Anthropology*, edited by George Stocking, 138–179. Madison: University of Wisconsin Press.

Pycraft, William P. 1925. 'On the Calvaria Found at Boskop, Transvaal, in 1913, and its Relationship to Cromagnard and Negroid Skulls'. *Journal of the Royal Anthropological Institute* 60: 179–198. https://doi.org/10.2307/2843700.

Rassool, Cirraj and Patricia Hayes. 2002. 'Science and the Spectacle: /Khanako's South Africa, 1936–1937'. In *Deep hiStories: Gender and Colonialism in Southern Africa*, edited by Wendy Woodward, Patricia Hayes and Gary Minkley, 117–161. Amsterdam: Rodopi.

Reich, David. 2018. 'How Genetics Is Changing Our Understanding of "Race"'. *New York Times Sunday Review*, 23 March. Accessed 19 July 2021. https://www.nytimes.com/2018/03/23/opinion/sunday/genetics-race.html.

Reich, David. 2019. *Who We Are and How We Got Here: Ancient DNA and the New Science of the Human Past*. Oxford: Oxford University Press.

Relethford, John H. 2009. 'Race and Global Patterns of Phenotypic Variation'. *American Journal of Physical Anthropology* 139, no. 1: 16–22. https://doi.org/10.1002/ajpa.20900.

Reubi, Serge. 2010. 'Eugène Pittard, un savant contre les intellectuels?' *Schweizerische Zeitschrift für Geschichte* 17: 38–48. https://doi.org/10.5169/seals-306566.

Reynolds, Vernon. 1988. 'Letter to the Editor'. *Anthropology Today* 4, no. 3: 25–26.

Rightmire, G. Phillip. 1970. 'Iron Age Skulls from Southern Africa Re-assessed by Multiple Discriminant Analysis'. *American Journal of Physical Anthropology* 33, no. 2: 147–168. https://doi.org/10.1002/ajpa.1330330203.

Roberts, Derek F. 1997. 'Joseph Sydney Weiner'. In *The Encyclopaedia of the History of Physical Anthropology*, edited by Frank Spencer, 1108–1109. New York: Garland.

Robins, Steven. 2016. *Letters of Stone: From Nazi Germany to South Africa*. Cape Town: Penguin.

Robins, Steven. 2020. 'The Boomerang: How Eugenics and Racial Science in the German Colonies Rebounded on Europe and the Rest of the World'. In *Fault Lines: A Primer on Race, Science and Society*, edited by Jonathan Jansen and Cyrill Walters, 41–54. Stellenbosch: African Sun Media.

Robinson, G.A. 1977. 'The Riddle of FitzSimons' "Tzitzikama Cave"'. *Koedoe* 20, no. 1: 95–99. https://doi.org/10.4102/koedoe.v20i1.933.

Rushton, J. Phillippe. 1995. *Race, Evolution and Behaviour: A Life History Perspective*. New Jersey: Transaction Books.

Salmons, Josephine E. 1925. 'The Bush and Bantu Mandibles'. *South African Journal of Science* 22: 470–479. Accessed 30 July 2021. https://hdl.handle.net/10520/AJA00382353_8597.

Schaffer, Gavin. 2007. '"Scientific" Racism Again? Reginald Gates, the *Mankind Quarterly* and the Question of "Race" in Science after the Second World War'. *Journal of American Studies* 41, no. 2: 253–278. http://dx.doi.org/10.1017/S0021875807003477.

Schauder, Donald E. 1963. 'The Anthropological Work of F.W. Fitzsimons in the Eastern Cape'. *South African Archaeological Bulletin* 18: 52–59. https://doi.org/10.2307/3887506.

Schlebusch, Carina M., Marlize Lombard and Himla Soodyall. 2013. 'mtDNA Control Region Variation Affirms Diversity and Deep Substructure in Populations from Southern Africa'. *BMC Evolutionary Biology* 13, no. 56. https://doi.org/10.1186/1471-2148-13-56.

Schlebusch, Carina M., Helena Malmström, Torsten Günther, Per Sjödin, Alexandra Coutinho, Hanna Edlund, Arielle R. Munters, Mario Vicente, Maryna Steyn, Himla Soodyall, Marlize Lombard and Mattias Jakobsson. 2017. 'Southern African Ancient Genomes Estimate Modern Human Divergence to 350,000 to 260,000 Years Ago'. *Science* 358: 652–655. https://doi.org/10.1126/science.aao6266.

Schramm, Katharina. 2016. 'Cast, Bones and DNA: Interrogating the Relationship between Science and Postcolonial Indigeneity in Contemporary South Africa'. *Anthropology Southern Africa* 39, no. 2: 131–144. https://doi.org/10.1080/23323256.2016.1168267.

Schramm, Katharina. 2020. 'Stuck in the Tearoom: Facial Reconstruction and Postapartheid Headache'. *American Anthropologist* 122, no. 2: 342–355. https://doi.org/10.1111/aman.13384.

Schwarz, Ernest H.L. 1919. *The Kalahari: Or Thirstland Redemption*. Cape Town: Maskew Miller.

Schwarz, Ernest H.L. 1923. 'The Kalahari Scheme as the Solution of the South African Drought Problem'. *South African Journal of Science* 20: 208–222. Accessed 30 July 2021. https://hdl.handle.net/10520/AJA00382353_8532.

Sealy, Judith C. and Susan Pfeiffer. 2000. 'Diet, Body Size and Landscape Use among Holocene Peoples in the Southern Cape, South Africa'. *Current Anthropology* 41, no. 4: 642–655. https://doi.org/10.1086/317392.

Sergi, Sergio. 1912. *Crania Habessinica: Contributo all'Antropologia dell'Africa Orientale*. Rome: Ermanno Loescher & Co.

Sharp, John. 1981. 'The Roots and Development of Volkekunde in South Africa'. *Journal of Southern African Studies* 8, no. 1: 16–36. https://doi.org/10.1080/03057078108708032.

Shaw, E. Margaret. 1978. 'Ethnology in Southern African Museums'. *South African Museums Association Bulletin* 3, no. 2: 1–17.

Shaw, E. Margaret. 1988. 'The Growth of a Collection'. *Sagittarius* 3, no. 4: 2.

Shear, Mervyn. 1996. *Wits: A University in the Apartheid Era*. Johannesburg: Wits University Press.

Shrubsall, Frank C. 1896. 'Crania from Tenerife'. *Proceedings of the Cambridge Philosophical Society* 9, no. 3.

Shrubsall, Frank C. 1898a. 'Crania of African Bush Races'. *Journal of the Royal Anthropological Institute* 27: 263–292. https://doi.org/10.2307/2842968.

Shrubsall, Frank C. 1898b. 'Notes on Ashanti Skulls and Crania'. *Journal of the Royal Anthropological Institute* 28, no. 1–2: 95–103. https://doi.org/10.2307/2842934.

Shrubsall, Frank C. 1898c. 'A Study of A-Bantu Skulls and Crania'. *Journal of the Royal Anthropological Institute* 28, no. 1–2: 55–94. https://doi.org/10.2307/2842933.

Shrubsall, Frank C. 1901. 'Notes on Crania from the Nile-Welle Watershed'. *Journal of the Royal Anthropological Institute* 31: 256–260. https://doi.org/10.2307/2842801.

Shrubsall, Frank C. 1902. 'Remarks on the Skeleton of a Bambute Pygmy from the Semliki Forest, Uganda Borderland'. In *The Uganda Protectorate*, by Sir H.H. Johnston, 559–565. London: Hutchinson & Co.

Shrubsall, Frank C. 1907. 'Notes on Some Bushman Crania and Bones from the South African Museum, Cape Town'. *Annals of the South African Museum* 5: 227–270.

Shrubsall, Frank C. 1911. 'A Note on Craniology of South African Bushmen'. *Annals of the South African Museum* 8: 202–208.

Shrubsall, Frank C. 1922. 'A Note on Bushman Craniology'. *Man* 22: 185–187.

Singer, Ronald. 1946. 'Some Aspects of Bushman Anthropology'. *Retina* 3, no. 2: 3–5.

Singer, Ronald. 1950. 'Discoveries in Southern Africa Contributing to Our Knowledge of the Ancestry of Man'. Cornwall and York Prize. Unpublished.

Singer, Ronald. 1953a. 'Estimation of Age from Cranial Suture Close: A Report of its Unreliability'. *Journal of Forensic Medicine* 1: 52–59.

Singer, Ronald. 1953b. 'The Sickle Cell Trait in Africa'. *American Anthropologist* 55, no. 6: 634–648. Accessed 30 July 2021. https://www.jstor.org/stable/664721.

Singer, Ronald. 1954. 'The Saldanha Skull from Hopefield, South Africa'. *American Journal of Physical Anthropology* 12, no. 3: 345–362. https://doi.org/10.1002/ajpa.1330120309.

Singer, Ronald. 1955. 'Race, Biology and Genetics'. In *Nusas Lecture Series on the University, Nationalism and Race*, 55–69. Cape Town: University of Cape Town.

Singer, Ronald. 1958. 'The Boskop "Race" Problem'. *Man* 58 (November): 173–178. https://doi.org/10.2307/2795854.

Singer, Ronald. 1993. 'Mythical African "Australoids" and Triangular Bricks: The Cape Flats Skull in Retrospect'. *South African Archaeological Bulletin* 48, no. 158: 105–112. https://doi.org/10.2307/3888950.

Singer, Ronald, Otto E. Budtz-Olsen, P. Brain and J. Saugrain. 1957. 'Physical Features, Sickling and Serology of the Malagasy of Madagascar'. *American Journal of Physical Anthropology* 15, no. 1: 91–124. https://doi.org/10.1002/ajpa.1330150113.

Singer, Ronald and Edward N. Keen. 1956a. 'Fossil Suiformes from Hopefield'. *Annals of the South African Museum* 42, no. 3: 149–179.

Singer, Ronald and Edward N. Keen. 1956b. 'Further Fossil Suidae from Hopefield'. *Annals of the South African Museum* 42, no. 4: 350–360.

Singer, Ronald and Joseph S. Weiner. 1963. 'Biological Aspects of Some Indigenous African Populations'. *Southwestern Journal of Anthropology* 19, no. 2: 168–176.

Singer, Ronald and John Wymer. 1968. 'Archaeological Investigations at the Saldanha Skull Site in South Africa'. *South African Archaeological Bulletin* 23, no. 91: 63–74. https://doi.org/10.2307/3888485.

Slater, Charles P. 1999. 'A Professor, a Student and a Criminal'. *South African Medical Journal* 89, no. 3: 254–256.

Slome, David. 1927. 'The Curvature of the Bushman Calvarium'. *American Journal of Physical Anthropology* 10, no. 3: 365–378. https://doi.org/10.1002/ajpa.1330100307.

Slome, David. 1929. 'The Osteology of a Bushman Tribe'. *Annals of the South African Museum* 35, no. 1: 33–60.

Slome, Isidore. 1932. 'The Bushman Brain'. *Journal of Anatomy* 67, no. 1: 47–58.

Smith, Grafton Elliot. 1917. 'Note on The Endocranial Cast Obtained from the Ancient Calvaria Found at Boskop, Transvaal'. *Transactions of the Royal Society of South Africa* 6, no. 1: 15–18. https://doi.org/10.1080/00359191709520169.

Smith, Grafton Elliot. 1923. *The Ancient Egyptians and the Origin of Civilization*. Revised edition. London: Cape.

Smuts, Jan Christian. 1926. *Holism and Evolution*. London: Macmillan.

Soudien, Crain. 2020. 'Knowing and Being: Living Our Learning about "Race"'. In *Persistence of Race*, edited by Nina G. Jablonski, 112–127. Stellenbosch: African Sun Media.

Spencer, Frank. 1997a. 'Haddon, A.C., 1855–1940.' In *The Encyclopaedia of the History of Physical Anthropology*, edited by Frank Spencer, 469–470. New York: Garland.

Spencer, Frank. 1997b. 'Sherwood L. Washburn'. In *The Encyclopaedia of the History of Physical Anthropology*, edited by Frank Spencer, 1104–1106. New York: Garland.

Spencer, Frank. 1997c. 'UNESCO Statement of Race'. In *The Encyclopaedia of the History of Physical Anthropology*, edited by Frank Spencer, 1052. New York: Garland.

Sperber, Geoff. 1990. 'Preface'. In *From Apes to Angels: Essays in Anthropology in Honor of Phillip V. Tobias*, edited by Geoff Sperber, ix–x. New York: Wiley-Liss.

Stepan, Nancy. 1982. *The Idea of Race in Science*. London: Macmillan.

Stone, Lee and Yvonne Erasmus. 2012. 'Race Thinking and the Law in Post-1994 South Africa'. *Transformation* 79, no. 1: 119–143. http://dx.doi.org/10.1353/trn.2012.0019.

Štrkalj, Goran. 2000. 'Inventing Races: Robert Broom's Research on the Khoisan'. *Annals of the Transvaal Museum* 37, no. 1: 113–124.

Štrkalj, Goran. 2005. 'A Note on the Early History of the Taung Discovery: Debunking the "Paperweight" Myth'. *Annals of the Transvaal Museum* 42, no. 1: 97–98. Accessed 19 July 2021. https://hdl.handle.net/10520/EJC83627.

Štrkalj, Goran. 2006a. 'Some Notes on the Early History of the Taung Discovery'. In *Collected Works for 'The 40th Anniversary of Yuanmou Man Discovery and the International Conference on Palaeoanthropological Studies'*, edited by Yang Decong, 255–259. Kunming: Yunnan Science and Technology Press.

Štrkalj, Goran. 2006b. 'Wells, Lawrence Herbert'. *S2A3 Biographical Database of Southern African Science*. Accessed 19 July 2021. http://www.s2a3.org.za/bio/Biograph_final.php?serial=3096.

Štrkalj, Goran. 2016. 'Raymond Arthur Dart's Academic Versatility'. *International Journal of History and Philosophy of Medicine* 6: 1–2. http://dx.doi.org/10.18550/ijhpm.0605.

Štrkalj, Goran and Phillip Tobias. 2008. 'Raymond Dart as a Pioneering Primatologist'. *Homo* 59, no. 4: 271–286. https://doi.org/10.1016/j.jchb.2008.06.002.

Štrkalj, Goran and A. Tracey Wilkinson. 2006. '"Race" and Biomedical Research: An Educational Perspective'. *Education for Health* 19, no. 1: 111–114. Accessed 19 July 2021. https://www.proquest.com/openview/cc6b1a652ddc7069bed24363c35ea053/1?pq-origsite=gscholar&cbl=33821.

Stuckenberg, Brian R. 2002. 'Obituary: John Adams Pringle: 1910–2002'. *Natalia* 32: 51–53. Accessed 19 July 2021. http://www.natalia.org.za/Files/32/Natalia%20v32%20Obits%20Pringle%20.pdf.

Stynder, Deano D., Rebecca Rogers Ackermann and Judith C. Sealy. 2007. 'Craniofacial Variation and Population Continuity during the South African Holocene'. *American Journal of Physical Anthropology* 134, no. 4: 489–500. https://doi.org/10.1002/ajpa.20696.

Summers, Roger F.H. 1975. *A History of the South African Museum, 1825–1975*. Cape Town: Balkema.

Swart, Andre L., P. Bernard Hesslig, Michael R. Hayden, Moesha Louw and J.S. Herbert. 1987. 'Gaucher's Disease in the Cape Coloured Population of the RSA, Including a Family with 5 Affected Siblings'. *South African Medical Journal* 71, no. 2: 97–99. Accessed 20 July 2021. https://journals.co.za/doi/pdf/10.10520/AJA20785135_7859.

Thomas, Harvey and Godfrey de Bruyn. 1985. 'Mentors'. *Era Magazine*, November: 4–5.

Thomson, Robert B. 1913. 'Note on the Vertebral Column of the Bushman Race of South Africa'. *Transactions of the Royal Society of South Africa* 3: 365–378. https://doi.org/10.1080/00359191309519704.

Tobias, Phillip V. 1953. 'The Problem of Race Identification: Limiting Factors in the Investigation of the South African Races'. *Journal of Forensic Medicine* 1, no. 2: 113–123.

Tobias, Phillip V. 1954. 'On a Bushman-European Hybrid Family'. *Man* 54: 179–182. https://doi.org/10.2307/2794001.

Tobias, Phillip V. 1955. 'Physical Anthropology and Somatic Origins of the Hottentots'. *African Studies* 14, no. 1: 1–15. https://doi.org/10.1080/00020185508706944.

Tobias, Phillip V. 1955/56. 'Les Bochimans Auen et Naron de Ghanzi: Contribution à l'étude des "Anciens Jaunes" Sud Africains'. *L'Anthropologie* 59: 235–253, 429–461; 60: 22–52, 268–280.

Tobias, Phillip V. 1957. 'The Bushmen of the Kalahari'. *Man* 57: 33–40.

Tobias, Phillip V. 1959. 'Some Developments in South African Physical Anthropology'. In *The Skeletal Remains of Bambandyanalo*, by Alexander Galloway, 129–154. Johannesburg: Wits University Press.

Tobias, Phillip V. 1961a. *The Meaning of Race*. Johannesburg: South African Institute of Race Relations.

Tobias, Phillip V. 1961b. 'The Work of the Gorilla Research Unit in Uganda'. *South African Journal of Science* 57, no. 11: 297–298. Accessed 30 July 2021. https://hdl.handle.net/10520/AJA00382353_6143.

Tobias, Phillip V. 1968. 'Alexander Galloway'. In *Southern African Dictionary of National Biography*, compiled by Eric Rosenthal, 284–285. London: Warne.

Tobias, Phillip V. 1970. 'Human Biological Research in the Anatomy Department, University of the Witwatersrand, 1959–1969'. *The Leech* 11, no. 2: 23–33.

Tobias, Phillip V. 1972. *The Meaning of Race*. 2nd edition, revised and enlarged. Johannesburg: South African Institute of Race Relations.

Tobias, Phillip V. 1974. 'The Biology of the Southern African Negro'. In *The Bantu-Speaking Peoples of Southern Africa*, edited by W.D. Hammond-Tooke, 3–45. London: Routledge and Kegan Paul.

Tobias, Phillip V. 1975. 'Fifteen Years of Study on the Kalahari Bushmen or San: A Brief History of the Kalahari Research Committee'. *South African Journal of Science* 71: 74–78. Accessed 20 July 2021. https://www.goandproclaim.co.za/downloads/Fifteen_Years_of_Study_on_the_Ka.pdf.

Tobias, Phillip V. 1984. *Dart, Taung and the 'Missing Link'*. Johannesburg: Wits University Press.

Tobias, Phillip V. 1985a. 'History of Physical Anthropology in Southern Africa'. *Yearbook of Physical Anthropology* 28, no. S6: 1–52. https://doi.org/10.1002/ajpa.1330280503.

Tobias, Phillip V. 1985b. 'Prehistory and Political Discrimination'. *South African Journal of Science* 81: 667–671. Accessed 30 July 2021. https://www.jstor.org/stable/41820816.

Tobias, Phillip V. 1987. 'Memories of Robert James Terry (1871–1966) and the Genesis of the Terry and Dart Collections of Human Skeletons'. *Adler Museum Bulletin* 13, no. 2: 31–34.

Tobias, Phillip V. 1988. 'Lawrence Herbert Wells (1908–1980) and the Wits Anatomy Department with Some Glimpses of His Role in the Australopithecine Unravelling'. *Adler Museum Bulletin* 14, no. 2: 11–17.

Tobias, Phillip V. 1990. 'The Role of R.B. Thomson and E.P. Stibbe: Brief Heralds of the Science of Anatomy in South Africa – Part I, R.B. Thomson'. *South African Medical Journal* 78, no. 6: 330–335, 426–430.

Tobias, Phillip V. 1991. 'Memories, Images and Visions: A Valedictory Address'. In *Images of Humanity: The Selected Writings of Phillip V. Tobias*, 1–26. Johannesburg: Ashanti Publishing.

Tobias, Phillip V. 1997. 'Some Little Known Chapters in the Early History of the Makapansgat Fossil Hominid Site'. *Palaeontologica Africana* 33: 67–79. Accessed 20 July 2021. http://wiredspace.wits.ac.za/bitstream/handle/10539/16402/1997. V33.TOBIAS.LITTLE%20KNOWN%20MAKAPANSGAT.pdf?sequence= 1&isAllowed=y.

Tobias, Phillip. 2000. 'Last Word: In Conversation with Phillip Tobias'. *Arena* 7, no. 23: 36–37.

Tobias, Phillip V. 2005. *Into the Past: A Memoir*. Johannesburg: Picador.

Tobias, Phillip V., Goran Štrkalj and Jane Dugard. 2008. *Tobias in Conversation: Genes, Fossils and Anthropology*. Johannesburg: Wits University Press.

Todd, T. Wingate 1937. 'The Scientific Influence of Sir Grafton Elliot Smith'. *American Anthropologist* 39, no. 3: 523–536. https://doi.org/10.1525/ aa.1937.39.3.02a00120.

Townsend, John. 2001. *Burke and Hare: The Body Snatchers*. Cheltenham: Nelson Thomas.

Trevor-Jones, Trevor R. 1980. 'Obituary: Professor L.H. Wells'. *South African Journal of Science* 76: 261–262.

Turner, A. Logan. 1933. *History of the University of Edinburgh, 1883–1933*. London: Oliver & Boyd.

Turner, M. 1970. 'A Search for the Tsitsikamma Shelters'. *South African Archaeological Bulletin* 25, no. 98: 67–70. https://doi.org/10.2307/3887954.

Ucko, Peter. 1987. *Academic Freedom and Apartheid: The Story of the World Archaeological Congress*. London: Duckworth & Co.

UCT (University of Cape Town). 2003. 'Teaching at UCT during Apartheid'. In *Truth and Reconciliation: A Process of Transformation at UCT Health Sciences Faculty*, 146–183. National Research Foundation Project Report 15/1/3/21/0022. Cape Town: University of Cape Town. Accessed 20 July 2021. http://www.cilt.uct. ac.za/usr/health/about/transformation/reconciliation/rec_ch5.pdf.

Underhill, David. 2011. 'A History of Stone Age Archaeological Study in South Africa'. *South African Archaeological Bulletin* 66, no. 193: 3–14. Accessed 20 July 2021. https://www.jstor.org/stable/41408528.

Unesco (United Nations Educational, Scientific and Cultural Organization). 1969. *Four Statements on the Race Question*. Paris: Unesco. http://refugeestudies. org/UNHCR/UNHCR.%20Four%20Statements%20on%20the%20Race%20 Question.pdf.

Van der Bank, Dirk Adriaan. 1995. 'Die Grey Universiteitskollege, 1904–1935'. *Navorsinge van die Nasionale Museum* 11, no. 11: 345–348. Accessed 30 July 2021. https://hdl.handle.net/10520/AJA00679208_1103.

Van der Bank, Dirk Adriaan. 1998. 'Die Van Hoepen Era, 1922–1950'. *Navorsinge van die Nasionale Museum* 14, no. 5: 97–143. Accessed 30 July 2021. https://hdl. handle.net/10520/AJA00679208_1743.

Van der Bank, Dirk Adriaan. 2000. 'A.C. Hoffman se Direkteurskap, 1951–1969'. *Navorsinge van die Nasionale Museum* 16, no. 5: 86–138. Accessed 30 July 2021. https://hdl.handle.net/10520/AJA00679208_854.

Van Hoepen, Egbert C.N. 1928. 'Die Koninge Kultuur 1. Die Koningse Industrie'. *Argeologiese Navorsing van die Nasionale Museum* 1, no. 1: 1–12.

Van Hoepen, Egbert C.N. and Abraham C. Hoffman. 1935. 'Die Oorblyfsels van Buispoort en Braklaagte, Noordwes van Zeerust'. *Argeologiese Navorsing van die Nasionale Museum* 2, no. 1: 1–25.

Van Niekerk, Anton A. 2011. 'Deliberating about Race as a Variable in Biomedical Research'. *South African Medical Journal* 101, no. 4: 248–250. https://doi.org/10.7196/samj.4702.

Van Pletzen, Rocco. 1987. 'Thomas Frederik Dreyer'. In *Dictionary of South African Biography*, vol. 5, edited by C.J. Beyer and J.L. Basson. Pretoria: Human Sciences Research Council.

Vargas, Luis A. 1997. 'Comas, Juan 1900–1979'. In *The Encyclopaedia of the History of Physical Anthropology*, edited by Frank Spencer, 290–292. New York: Garland.

Vedder, Heinrich. 1938. *South West Africa in Early Times*. Translated by Cyril G. Hall. London: Oxford University Press.

Vermooten, Vincent. 1921. 'A Study of the Fracture of the Epistropheus due to Hanging with a Note on the Possible Causes of Death'. *The Anatomical Record* 20, no. 3: 305–311. https://doi.org/10.1002/ar.1090200305.

Vigilant, Linda, Mark Stoneking, Henry Harpending, Kristen Hawkes and Allan C. Wilson. 1991. 'African Populations and the Evolution of Human Mitochondrial DNA'. *Science* 253, no. 5027: 1503–1507. Accessed 20 July 2021. https://www.jstor.org/stable/2884983.

Von Luschan, Felix. 1906. 'Bericht uber eine Reise in Sudafrika'. *Zeitschrift für Ethnologie* 38: S863–895.

Von Luschan, Felix. 1907. *The Racial Affinities of the Hottentots*. Report of the British and South African Associations. London: Spottiswoode.

Walters, Handri. 2018. 'Tracing Objects of Measurement: Locating Intersections of Race, Science and Politics at Stellenbosch University'. PhD diss., University of Stellenbosch.

Washburn, Sherwood L. 1951. 'The New Physical Anthropology'. *Transactions of the New York Academy of Sciences* 13, no. 7: 298–304. https://doi.org/10.1111/j.2164-0947.1951.tb01033.x.

Weiner, Joseph S. 1952. 'Physical Anthropology since 1935: A Survey of Developments'. In *A Hundred Years of Anthropology*, 2nd edition, edited by T. Kenneth Penniman, 376–412. London: Duckworth & Co.

Weiss, K.M. and T. Maruyama. 1976. 'Archeology, Population Genetics and Studies of Human Racial Ancestry'. *American Journal of Physical Anthropology* 44: 31–50. https://doi.org/10.1002/ajpa.1330440106.

Wells, Lawrence H. 1931. 'The Foot of the South African Native'. *American Journal of Physical Anthropology* 25, no. 2: 185–289. https://doi.org/10.1002/ajpa.1330150202.

Wells, Lawrence H. 1952. 'Physical Measurements of Northern Bushmen'. *Man* 52: 53–56. https://doi.org/10.2307/2795077.

Welsh, David. 2009. *The Rise and Fall of Apartheid*. Johannesburg: Jonathan Ball.

West, Martin. 1988. 'Confusing Categories: Population Groups, National States and Citizenship'. In *South African Keywords: The Uses and Abuses of Political*

Concepts, edited by Emile Boonzaier and John Sharp, 100–110. Cape Town: David Philip.

Wheelhouse, Frances. 1983. *Raymond Arthur Dart: A Pictorial Profile.* Sydney: Transpareon Press.

Wheelhouse, Frances and Kathaleen S. Smithford. 2001. *Dart: Scientist and Man of Grit.* Sydney: Transpareon Press.

Wickham Legg, Leopold G. 1949. 'Grafton Elliot Smith'. In *Dictionary of National Biography, 1931–1940.* Oxford: Oxford University Press.

Wilkinson, Caroline. 2010. 'Facial Reconstruction: Anatomical Art or Artistic Anatomy?' *Journal of Anatomy* 216: 235–250. https://dx.doi.org/10.1111%2Fj.1469-7580.2009.01182.x.

Wilson, Michael and William J.J. van Rijssen. 1990. 'The Coldstream Stone: A Unique Artefact in the South African Museum Collections'. *Sagittarius* 5, no. 1: 6–8.

Wood, Bernard. 1989. 'An Interview with Phillip Tobias'. *Current Anthropology* 30, no. 2: 215–224. https://doi.org/10.1086/203735.

Worboys, Michael. 1981. 'The British Association and Empire: Science and Social Imperialism'. In *The Parliament of Science,* edited by R. MacLeod and P. Collins, 170–189. London: Science Reviews.

Yap, Melanie and Dianne L. Man. 1996. *Colour, Confusion and Concessions: The History of the Chinese in South Africa.* Hong Kong: Hong Kong University Press.

Ziegelmayer, G. 1997. 'Otto Schlaginhaufen'. In *The Encyclopaedia of the History of Physical Anthropology,* edited by Frank Spencer, 912. New York: Garland.

Zuckerman, Solly. 1926. 'Growth-Changes in the Skull of the Baboon, *Papio Porcarius'. Proceedings of the Zoological Society of London* 96, no. 3: 843–873. https://doi.org/10.1111/j.1469-7998.1926.tb07131.x.

Zuckerman, Solly. 1970. *Beyond the Ivory Tower.* London: Weidenfeld and Nicholson.

Zuckerman, Solly. 1978. *From Apes to Warlords.* London: Hamish Hamilton.

INDEX

Page numbers in *italics* indicate figures or tables.

#BlackLivesMatter 293
!Kung San xix, 175, 212
!Ora Khoekhoe xix
/Keri-/Keri 182
/Khanako 90
/Xam San xix
‡Khomani San xix, 235

A
academic boycott 241–243
Acheulian 12–13
Adesina, Jimi 292–293
African Expedition, University of
    California 175
African National Congress (ANC)
    242, 272
Afrikaans language 134–135
Afrikaans-language anthropology 18,
    132–142
Afrikaans speakers 15–16, 248, 268, 269
Afrikaner nationalism 134–135, 248,
    268–269
Aging, Neuropsychology, and Cognition
    291
Albany Museum, Grahamstown 7, 26
Alexander, Neville 5
Allbrook, David 233
American Association of Physical
    Anthropologists 283–284
*American Journal of Physical
    Anthropology* 172

American Museum of Natural History
    131, 160
American National Institutes of Health
    217–218
America, typological views of race in 294
Ames, Francis 240
Anatomical Record 215
anatomical schools, private 67, 69
anatomy 2–3, 66–70, 76–77, 140, 149, 236
Anatomy Act (Britain) 69
Anatomy Museum, Edinburgh University
    70–71
ANC *see* African National Congress
Andrews, H.T. 35
Angel, J. Lawrence 206
Angola, expedition to 126–129, 141, 175
Annals of the South African Museum
    23, 85
Anthropology Today 279
anthropometry 137–138, 173–174
apartheid
    Afrikaans-language anthropology
        and 18
    Dreyer on 134
    ethnos theory 267–268
    history of 247–248
    lawyers and 265
    legacy of 1–2, 285, 294–295
    petty apartheid 264–265
    physical anthropologists and 247,
        267–274

apartheid (*continued*)
    race classification 2–3, 5–6, 8–9,
        179–180, 248–268, *251*, 2
        85–289, 293
    race reclassification 252–266,
        254–255, 258, 259, 261
    Singer on 216–217
    Tobias on 196, 224, 226, 239–243
archetype 109–110
*Argeologiese Navorsing van die Nasionale
    Museum* 125, 134–135
armed forces 66–67, 69
Arnold, Peter 182
Association of Medical Students of South
    Africa 182
Association of Southern African
    Archaeologists 243
atrocities, Nazi regime 219–220
Auen (=Au//ei) San 202
Aurignacian 13, 14, 39
Australoids xix, 132

B
baboon research 150, 232
Badenhorst, T.R. *238*
Bahurutsi people 125
Bain, Donald 89–90, 170
Balsan, François 199–200, *201*, 202
Bambandyanalo site 55, 57, 142, 168–169,
    169, 173
Bantu-speaking peoples xix, 23, 24, 126,
    173, 235
Bantu Studies 170
Barlow, G.W. *198*
Barnard, Keppel *19*, 36
Barnard, Tom 148
Barry, Marie 112, 115, 134
Barry, Tom 218
Baumgartel, Walter 232–233
behaviour and biology, link between
    133, 141
behaviour and race, link between 17–18,
    133, 186
Belgian Congo, trip to 157–158
beliefs and science 294
Bibescu, Prince George 27

Biko, Steve, death of 239–240
biological distance 276, 278, 281
biology and behaviour, link between
    133, 141
biology and culture 249, 284–285, 288
biology as subject, university level
    292–293
biometrical analysis 204–205, 220–221,
    275–276
Black Consciousness organisations
    272, 273
Bleek, Dorothea 83–84, 132
Bloemhof, expedition to 173–174
Boers *see* Afrikaans speakers
Boer War *see* South African War
Bolk, Lodewijk (Louis) 56, 89
Bolwig, Niels 233
Boskop skull
    analysis of 43–49, 110
    discovery of 27, 40–43, *41*
    reports on 48–63, 82
Boskop type xix, 50–63, 79, 83, 112, 168,
    210–213, 223, 230
Botha, Piet 40–43
Boule, Marcellin 137
    Hommes fossiles, Les 137
boycotts
    academic 241–243
    school 258–260
Boyd, William C. 206
brain size and function 48, 52, 53,
    60, 110
Breakwater Prison, San men from 18–19
Brenner, Sydney 93, 195
Brew, John O. 206–207
Bridgman Memorial Maternity Hospital
    174, 181–182
Brink, Arend 121
Brink, Vernon Hofmeyr xvii, 31,
    121–123, *122*, 137, 139–140
Britain 16, 69, 294
British Association of Science 15–16, 18
Brit, unity between Boer and 15, 16
Broad-Based Black Economic
    Empowerment Act 286
Broca, Paul 27

Broken Hill Development Company 157
Broom, Robert
    background of xvi
    Boskop skull 48–49, 60–61
    craniology 121, 132
    in Douglas 130–131
    Drennan and 78
    Dreyer and 132–133
    eugenics movement 140
    Grobbelaar and 136
    Kakamas type 133
    Khoesan peoples 131–132
    Neanderthaloid individuals 88
    Riet River graves 130–132
    Springbok Flats 59
    Sterkfontein fossil site 198
    Tobias and 197
    at Victoria College 111, 130
Budtz-Olsen, Otto 212
Bulawayo Chronicle 181
Bureau of Census and Statistics 251
Burke and Hare scandal (1829) 68–69
Burke, William 69
Burns, Catherine 174, 235

C
cairns 121, 133
Cairo Museum 160
Cambridge medical school 68–69
Campbell Griqua community 237–238,
    238
Canadian Association of Science 16
Cape Action League 273
Cape Argus 59
Cape Colony 3, 247–248
Cape Flats, bones found in 86–90, 87
Cape Khoe xix, 26
Cape Times 50, 58, 59, 270
Cape Town Medical School 73–74
Carr-Saunders, Alexander 220
    We Europeans 220
casts 17, 18–19, 84, 155–156, 175,
    201–202
Caton-Thompson, Gertrude
    154–155, 167
cave art 52 see also rock art

cave sites 20–21, 23–24, 26, 49–55, 61, 62
Centre for African Studies, UCT 4–5
Challenger Expedition 70
Chaplin, George 284
Chicago Tribune 215
Chinese, race classification of 257
Christiana, expedition to 173–174
Churchill, Winston 1
Cipriani, Lidio xvii, 155–157, 183–184,
    202
Clarke, Ron 240
Clement, A.J. 233
Cobern Street, Cape Town, burials from
    289–290
'coefficient of racial likeness' 275–276
Coetzee, Daan 214
Cohen, Solly 84–85
Coldstream Cave 20–21
Colesberg cemetery 20, 84, 85–86
colonised education 293
coloured population 5–6, 138–139,
    249–250, 291–292
Comas, Juan 33–34, 220–221
Congo, trip to 157–158
Coon, Carleton 206
Cornwall and York Prize 93, 209–210
Correia, Janine 65–66
Corstorphine, George Steuart 12
courts 277
Crail, Archibald 272–273
craniology 22–26, 32–33, 110, 121, 170,
    178–179
Cro-Magnon people xix, 13, 39, 48,
    52, 53
culture and biology 249, 284–285, 288
culture and race, link between 18
Cunningham, Andrew 66–67
Cunningham, Daniel John 71–72
Cuvier, Georges 68

D
Dahlberg, Gunnar 220
    Race, Reason and Rubbish 220
Dart, Raymond
    background of xvii, 144–145
    biometrical analysis 204

Dart, Raymond (*continued*)
  Boskop type xix, 50–52, 54, 58,
    83, 168
  craniology 121
  De Villiers and 231
  diffusionism 165, 167–168, 179–184
  Drennan and 82–83
  eugenics movement 140
  Galloway and 178, 186
  Gates and 184–186
  gorilla research 232–233
  Great Zimbabwe ruins 154–155, 157
  health of 171–172
  Italian scientific expedition
    155–161, 202
  Kalahari expeditions 170, 235
  pedomorphs xix, 56
  political views 181–184, 269, 270
  publications of 165, 171–172
  race classification 263, 267
  research interests 29
  Singer and 208, 213
  Taung skull 58, 150–155, *151*, 160,
    162–163, 181, 183, 186
  Tobias and 194, 197–200, 203, 219,
    224–226, *240*, 269
  typology 167, 174, 179–184
  at University College, London 144
  University of Birmingham and
    163–164
  Wells and 177–178, 186
  at Wits 50–51, 137–138, 145–155,
    164–165
Davie, T.B. 101
Day, John 218
Deepwalls Research Station 148
'degenerate' 127
Del Grande, Nino 156–157, 159–162,
  183
Dendy, Arthur 111, 112
Department of Anatomy, UCT 7–8,
  79, 208
Department of Anatomy, Wits 47
Department of Human Biology, UCT 6–8
Department of Internal Affairs 253, 256
Derricourt, Robin 183, 269

De Villiers, Cornelis Gerhardus
    Stephanus (Con) xvii, 137,
    139–141, 269, 271
De Villiers, Hertha
  background of xvii, 230–231
  craniology 236, 276
  Dart and 171, 231
  Kalahari expeditions 234–235
  race classification 264–265
  Singer and 208
  Tobias and 204, 227, 229–230, 232,
    237–239
  at Wits 214, 231–232
diffusionism 165–168, 170, 179–184,
    268–269
Discover magazine 60, 62
Drennan, Matthew Robertson
  anthropology, interest in 79–82
  background of xvi, 73–75
  Boskop type 54, 56, 59, 79
  Cape Flats, bones found in 86–90
  craniology 121
  Dreyer and 116, 117
  at Edinburgh 65, 72
  Elandsfontein site 95–102, *100*
  eugenics movement 140
  Grobbelaar and 138
  influences on 82–85, 90–95
  legacy of 102–103
  political views 270–272
  publications of 85–90
  research interests 75–83
  Short Course on Human Embryology,
    A 80
  Singer and 94, 208, 209, 210, 212
  typology 86, 123, 138, 142
  at UCT 73–82, *73*, 94–95, 103, 213
Dreyer, Thomas Frederick xvi, 5–6, 8,
    111–121, *113*, *118*, 124, 126–129,
    132–141
Drury, James xvi, *19*, 20–21, 47, 50,
    84–86, 123–124, 148
Dry, Thomas 84–85
Dubow, Saul 53, 111, 180–181, 248,
    269
Duckworth, Wynfrid L.H. 22, 150

E

Early Stone Age 14
Eastern Province Herald 58
East India Company 69
Edgar, Heather 294
Edinburgh 66–73, 79
Edinburgh Medical School 70–71, 271
Edinburgh University 69–73
education *see* medical education; school
　　boycotts
Egypt 158, 165–166, 179–180
Elandsfontein site 95–102, *95*, *96*, *98*, 210,
　　213, 217
elections 248, 257–258
electrocardiograph data collection 234
Ellison, George 261–263, 286–287
embryology 80–81, 210
Engelbrecht, J. 119
Erasmus, Yvonne 256–263
ethnographic photography 156
ethnos theory 17–18, 267–268
eugenics movement 140–141, 204,
　　219–220, 268–269
Eugenics Society of South Africa 140–141
Eurocentric vision 292
evolution 2, 284
Extension of University Education Bill
　　213–215, 239

F

face masks/moulds 156, 175, 201–202
facial reconstructions 289–290
faith and science 294
Farini, Gilarmi 199–200
Farmer's Weekly 136
Fascist Institute of Culture 160
Findlay, George 111, 130
Fischer, Eugen 137, 139–140
Fisher, Ronald 220
Fish Hoek Cave 54–55, 89, 218
FitzSimons, Frederick William xvi,
　　40–52, *44*, *49*, 54–55, 58, 61–62,
　　147, 149, 172
　　Monkey Folk of South Africa, The 46
Florisbad hominin 117–118, *118*,
　　126, 129

Fock, Gerhard *238*
Fordisc 276–277
Fouché, Leo 169
founder effect 284–285
Fourie, Louis 149
Fowler, William 130
franchise 248
Frobenius, Leo 154
functionalist school of social
　　anthropology 166
Fynn, Henry Francis 250

G

Galloway, Alexander (Sandy)
　　background of xvii
　　Bambandyanalo site 55, 142, 168,
　　　169, 173
　　Boskop type 54–55
　　Dart and 147, 171–172, 178, 186
　　eugenics movement 140
　　at Makarere University 178, 233
　　Mapungubwe site 168–169
　　typology 121, 142, 179
　　at Wits 160–162, 178
Garabedian, Star *19*
Gates, Reginald Ruggles 140, 183,
　　184–186
　　Human Ancestry 185
Gatti, Attilio 155, 156–159
Gaucher disease study 5–6
Gear, Harry Sutherland 50–52, 149
Gear, Jock 172
genetic drift 284–285
genetics 2, 281–282, 284–285
genitalia 84, 235–236
genome analysis 281–283
geographic variation 279–280, 282–284
geometric morphometrics 276
German approach to racial biology
　　139–140
German ethnos theory 17–18
gerontomorphs 89
Ger, Ralph 215–216
giganto-pedomorphic 56
Gilbert, Christine 194
Gill, Leonard 34–35, 132

Gillman, Joseph 194
Gill, Sir David 15
Gilman, Stanley *19*
Goodsir, John 70
Goodwin, John 14, 34–35, 54, 90, 96,
    99–102, 119, 134, 148
Gorfinkel, Max 84
gorilla research 232–233
Granger, Richard 60–63
    Big Brain 60–63
Grant, J.C.B. 80
Great Chain of Being 108–109
Great Zimbabwe ruins 154–155, 157
Greenberg, Jacquie 288
Grine, Fred *240*
Griqua people 31, 125–126, 139–141,
    235, 237–238, *238*, 257
Grobbelaar, Coert Smit xvi, 111,
    136–137, *138*, 269
Groenewald, J.L. 40–42
Group Areas Act 249, 250, 251, 257,
    266–267
Guenther, Lars 293–294

H
Haddon, Alfred Cort xvi, 16–18, 21–22,
    110–111, 220
    We Europeans 220
Haffajee, Ferial 292
Haldane, J.B.S. 204
Hall, Martin 5, 273
Hamitic theory 17, 22
Hammond-Tooke, W.D. (David) 224–225
Hare, William 69
Harrison, Geoffrey Ainsworth 217
Haughton, Sidney *19*, 43, 44, 48, 50
Hawks, John 63
Hayes, Patricia 170
Headlight, The 157
health research 286–289
Heikum San 212
Henderson, William Ramsay 71–72
Hendy, Brett 218
Heriot Bursary 76
Heselson, Jack 90–91
Hewitt, John 116–117

Hoernlé, Winifred 119, 148, 153,
    167–168
Hoffman, Abraham xvii, 125–129,
    136–141
Hofmeyr, Jan Hendrik 146, 153
Hogben, Lancelot 220
Homo capensis 48
Homo helmi 117
Hooper, Otto 84
Horrell, Muriel 253, 256
Howells, William (Bill) 57, 206–207
Hughes, Alun xvii, 5–9, 175, 182, 218, 227
human biology 223
human skeleton collecting 14–15, 19–21,
    68, 121, 124, 129–132
Humphreys, Anthony 130
Humphreys, Basil 237
Hunley, Keith 294
Huxley, Julian 220
    Evolution: The Modern Synthesis 220
    We Europeans 220

I
IBP *see* International Biological
    Programme
Illustrated London News 59, 82, 157
Immorality Act 248, 263–264
Immorality Amendment Act 248–249
Indians, race classification of 250–251,
    256–257, 260, 272
industrial development 16
inequality 285, 287, 294–295
insensitivity of scientists 235, 291–292
Inskeep, R.R. (Ray) 102
institutional racism 293
intelligence 60, 110, 291–292
International Biological Programme
    (IBP) 217, 223, 234–235
International Union of Prehistoric and
    Protohistoric Sciences (IUPPS)
    241–242
Inter-University Committee for African
    Studies 119
Italian scientific expedition 155–162
Italian system of craniology 170
Italy 184

ithyphally 235–236
IUPPS *see* International Union of Prehistoric and Protohistoric Sciences
Iziko Museum 7

J
Jablonski, Nina 284
Jantz, Richard 276–277
Japanese Asians, race classification of 250
Jenkins, Trefor 206, 227, 240, 263, 265–267, 279
  Peoples of Southern Africa and Their Affinities, The 279
Jenson, Joseph Stokes 197
Jewish refugees 140
Johannesburg Star 157
Jolly, Aimée 94–95
Jolly, Keith xvii, 59, *96*, 97, 99–102
Jolly, William 77, 95
Journal of the Royal Anthropological Institute 86, 88

K
Kakamas xix, 127, *128*, 133–134
Kalahari expeditions 156, 170, 199, 207, 233–234
Kalahari Research Committee (KRC) 207, 233–234
Kalahari San xix, 23, 26, 89–90, 156
Kalahari Thirstland Redemption Society 135–136
Kanam site, Kenya 205
Karoo 8, 120–124
Kaufman, Pamela De Beer 236
Kaufman, Tad *238*
Keen, E.N. (Ted) xvii, 5–8, 36, 91–95, 97–99, *98*, 101–102, 263–264
Keen, J.A. (Jack) xvii, 36, 91–93, 209
Keith, Arthur xvi, 23–24, 48, 53, 56, 88–89, 145, 150, 162–163
  Antiquity of Man, The 53
Keough, Natalie 118–119
Khoekhoe ('Hottentots') xix, 3, 17, 23–27, 30–31, 50, 78, 126–127, 133, 257

Khoesan peoples
  Boskop type and 57
  Broom, Robert 131–132
  Dart, Raymond 149
  Drennan, Matthew Robertson 56, 82–84, 86, 89–90
  Dreyer, Thomas Frederick 121, 132
  Drury, James 84
  first modern Europeans and 14
  genetic diversity of 282
  Péringuey, Louis 24
  Pittard, Eugène 32
  Singer, Ronald 223–224
  Vermooten, Vincent 77
  Wells, Lawrence 173–174
Kirsch, Ralph 175, 177
Kirsten, Johan Ferdinand van Eyck 214
Klasies River site 217–218
Knox, Robert 68–69
Kok, Adam, IV 237
Kok, Cornelis, II 237
Korana people xix, 31, 131–132, 173–174
Korean War Project 206
Krantshoek cave 148
KRC *see* Kalahari Research Committee
Kuljian, Christa 170, 182, 237, 264–265

L
L'Abbé, Ericka 118–119
Lady Buxton Children's Emergency Home 264
Laing, Gordon D. 50–53, 146–147
Lamarck, Jean-Baptiste 68
Land Act (1913) 248
Langebaanweg site 217–218
Lang, Ivor 240
Late Iron Age 273
Later Stone Age 14, *73*
lawyers 265
League of Nations 35
Leakey, Louis 205, 232
Leakey, Mary 232
Le Grange, Lesley 291–292
Le Gros Clark, Wilfrid 205, 222
Le Moustier site 13, 39
Lewontin, Richard 280–281

Lightfoot, Robert *19*
Linnaeus, Carl 108–109
   Systema Naturae 109
Lockhart, Robert Douglas 164
looking-glass similarity 81
Loots, Marius 118–119
Loram Committee 152, 181
Louw, J.T. 116
lumpers 273
Lurie, David 84
Lynch, Gary 60–63
   Big Brain 60–63

M
Mabbutt, Jack *96*, 97, 99
Macmillan, William Miller 153
macronymphae 235–236
Madagascar, expedition to 212
Maingard, J.F. 173
Maingard, Louis 173
Makapansgat site 197–199
Malan, Berry 197
Malherbe, D.F. 115–116
Malinowski, Bronisław 166–167
Manifesto on Race 184
Man in Africa exhibition 263
Mankind Quarterly 186
Manouvrier, Léonce 27
Mantegazza, Paolo 156
Mapungubwe site 55, 57, 168–169
Marks, Jonathan 2
   Why I Am Not a Scientist: Anthropology
     and Modern Knowledge 2
Marshall, Laurence and Lorna 207, 233
Martin, Rudolf 83, 137
   Lehrbuch der Antropologie 137
Mason, Revil 228
Matjes River site 54–55, 115–121, 126,
   128, 129
Mauduit, Jacques 200
Mazel, Aron 271
McGregor Museum, Kimberley 3, 26,
   124, 129–132
McKee, Jeff 232
McKittrick, Meredith 136
McManus, Gerald 86

media *see* newspapers, role of
medical education 29, 66, 68–72
Medical School, South African College
   21, 25, 29, 65, 76–77
Meiring, Albert (Bertie) John Dirk xvii,
   115, 119–120, 127–129, 133–141
Mendelian genetics 136, 139–140, 184,
   220, 223–224
mental abilities *see* intelligence
Middle Palaeolithic 13
Middle Stone Age 14, 54, 59
Midgley, Jeremy 293
mirror image homology 81
Mixed Marriages Act 248–249, 265
mixed racial groups 3, 136, 141, 269
Mnguni, Lindelani 293
molecular genetics 281–283
Monro, Alexander (*primus*) 67
Monro, Alexander (*secundus*) 67
Monro, Alexander (*tertius*) 67–68, 70
Montagu, Ashley 184–185, 220–221, 273, 279
   Man's Most Dangerous Myth 220
Morris, Alan G. 3–7, *240*, 266–267, 276,
   279–280, 288–290
   Master Catalogue, A: Holocene Human
     Skeletons from South Africa 4–5
   'Role of Physical Anthropology in
     the Study of African History, The'
     (paper) 5
Mosothwane, Morongwa 131
moulds *see* casts
Mousterian 13
Movius, Hallum 206
multiple gene hypothesis 220
multivariate statistical analyses 276–277
Mumbwa caves 157–158
mummification 166
Murray, Bruce 152
Mussolini, Benito 162

N
names, ethnic or population group
   288–289
Namibia see South West Africa
Naron (Nharo) San 202
nationalism 27, 132–142

National Museum, Bloemfontein 3, 7, 124–126
National Party 247, 248, 257–258, 271, 272
national types 110
National Union of South African Students (Nusas) 195–196
Nattrass, Nicoli 292–293
Natural History Museum, London 205
Natural History Museum of Milan 160
natural selection 278, 281–282, 284
Nature 150
Ncayiyana, Daniel 287, 289
Neanderthals xix, 39, 48, 53, 88, 89
Neel, James V. 206
'Negroids' xix, 23, 40
'new physical anthropology' 8–9, 57, 206, 221–225, 267, 275
newspapers, role of 58–59, 154
New York Times 157
Nieuwoudt, Sharné 291–292
nomenclature in archaeology 13–14, 44–45, 134
Nuffield Senior Fellowship 203–205
Nurse, George xvii, 237–238, 265–266, 279
Nusas *see* National Union of South African Students

O
Oakhurst site 90
Odendaal, Maria 122
Olmesdahl, P.J. *238*
Orford, Margaret 174, 235
Orkin, Carol 235
Oudtshoorn site 117
Ousley, Stephen 276–277
Oxford medical school 68–69

P
paedomorphs 89
palaeoanthropology 39–40, 48, 232
Panhard-Capricorn Expedition 199–203, *201*, 233
Paver, F.R. 58
Pearson, Karl 204, 275–276
pedomorphs xix, 56

Penrose calculations 230, 276
Péringuey, Louis
  archaeology, interest in 12–14
  background of xvi, 11–12
  Boskop type 42–44, 48
  casts of San men 18–19
  death of 30–32
  Dreyer and 115
  Drury and 84
  FitzSimons and 50
  human skeleton collecting 14–15, 20–21
  Pittard and 26–32
  S2A3 14, 18
  Shrubsall and 21, 22–26
  at South African Museum 11–12, *19*
  typology 78
  Vermooten and 77–78
Perry, William J. 166
phenetic analysis 281
Philippolis 125–126, 141
Phillips, John 148
photography, ethnographic 156
Piltdown skull 40, 85
Pittard, Eugène xvi, 7, 26–36, 78, 83, 123
Plato 109
Pöch, Rudolf 24
population genetics 220, 222
population register 285
Population Registration Act 249–253, *251*, 260
Port Elizabeth Museum 40–43, 46–47, 61
poverty 248, 269, 285, 287, 294–295
Princess Alice Adoption Home 264
Pringle, John 47–48
progressive scientists 273
Prohibition of Mixed Marriages Act 248, 265
public audiences 181, 284, 291
'pure races' 27, 109–110, 123, 127, 142
Pycraft, William 53, 110
pygmaeo-pedomorphic 56

Q
Qoboza, Percy 228
Quota Bill 239
quotas to redress past imbalances 286

R

race
  behaviour and 17–18, 133, 186
  classification 2–3, 9, 179–180,
    248–268, *251*, 285–289, 293
  craniology 132, 276–277
  culture and 18
  existence of 5, 223
  identity 276–277, 290
  intelligence and 110
  in post-1994 South Africa 285–295
  'pure races' 27, 109–110, 123, 127, 142
  reclassification 252–266, *254–255*,
    258, *259*, *261*
  relations surveys 253, 256
  species and 109
  views on 27, 33–34, 39–40, 109, 139,
    155, 224, 290–291
Race Classification Board 252–253, 260,
  262, 265–267
racial segregation *see* segregation
racial typology *see* typology
racism 1–2, 53–55, 110, 135, 140,
  180–184, 268, 293
Radcliffe-Brown, Alfred 166, 167
Raikes, Humphrey 160
Rand Daily Mail 58
Rassool, Ciraj 170
redress of past imbalances 285–289
'Rehoboth Basters' 139, 212
Reich, David 283, 285
Relethford, John 283
religion 68–69
resurrectionists 69
Retina 209
Reunert, Theodore 15
Reynolds, Vernon 279
Rhodesian skull 132
Riet River graves 129–132
Rightmire, G. Phillip 57, 276
Rivers, William H.R. 166
Robinson, Arthur 71–72, 81–82
Robins, Steven 291
rock art 13, 52, 53, 61
Rockefeller Fellowship 145, 203, 205–207
Romer, Alfred 206

Rotary Foundation Fellowship 93–94,
  210–211
Rushton, Phillippe 1

S

S2A3 *see* South African Association for
  the Advancement of Science
Saint-Hilaire, Étienne Geoffroy 68
Saldanha skull 59, 97–102, *97*, 213, 218
Salmons, Josephine 149, 150
SAMDC *see* South African Medical and
  Dental Council
SAMJ see *South African Medical Journal*
San people
  !Kung San xix, 175, 212
  /Xam San xix
  ‡Khomani San xix, 235
  Bleek, Dorothea 83–84
  Broom, Robert 49, 78
  Dart, Raymond 54, 170, 182
  De Villiers, Hertha 234
  Drennan, Matthew Robertson 89
  Drury, James 21, 84, 85–86, 123–124
  FitzSimons, Frederick William 50–51
  Hoffman, Abraham 126–127, 129
  Panhard-Capricorn Expedition
    201–203
  Péringuey, Louis 13, 18–20, 23–24,
    30–32
  racial overtones in descriptions 53
  rock art 61
  schema of types xix
  Singer, Ronald 209
  Tobias, Phillip 201–203
  United Empire Exhibition 89–90
  Von Luschan, Felix 17
  Wells, Lawrence 175
Santayana, George 1
Saunders, Peter 218
Schaller, George 233
Schapera, Isaac 224
  *Bantu-Speaking Peoples of Southern
    Africa, The* 225
  *Bantu-Speaking Tribes of South Africa,
    The* 224–225
Scher, Sam 81

school boycotts 258–260
school museums 45
School of African Life and Languages,
    UCT 167
Schramm, Katharina 289–291
Schwarz, Ernest 135–136
science, role in Union of South Africa
    15–17
scientists
    insensitivity of 291–292
    race classification 260–267, 280–285
Scottish influence in Cape Town
    66–73, 79
Sechaba 272
Second World War 35–36, 208, 219–220
sectioning point 277
Seftel, Harry 195
Seggie, Janet 288
segregation 214–216, 247–248, 285
self-defined names 289
Separate Amenities Act 249, 250, 251, 257
Sergi, Giuseppe 156
Sergi, Sergio 179
serial homology 81
serogenetics 278–281
Shaw, Margaret 5–7, 36
Shellshear, Joseph 145
Shore, Lewis R. 147
Shrubsall, Frank C. xvi, 7, 21–26, 78, 83
Singer, Martin 207
Singer, Ronald
    anthropology, interest in 93, 210–212,
        218
    background of xvii, 207–208
    Boskop type 56–57, 210, 211–212,
        213, 223–224
    Dart and 226
    Drennan and 102
    Elandsfontein site *96*, 97–102, 213
    in England 211–212
    Keen and 92–93
    'new physical anthropology' 57, 186,
        192, 223
    papers of 218–219
    personality of 218
    political views 214–217, 219, 266

research interests 209–210
Rotary Foundation Fellowship 93–94,
    210–211
Saldanha skull 59
South West Africa research 217
Tobias and 219
at UCT 86, 94–95, 208–210, 212–216
at University of Chicago 216–218
Weiner and 223
Wells and 177, 179
Singer, Solomon 207
skeleton collecting *see* human skeleton
    collecting
Slome, David 85–86, 123
smallpox epidemic (1866) 85, 123
Smithford, Kathaleen 157
Smith, Grafton Elliot xvi, 43, 48, 144–
    145, 150, 162–163, 165–168, 179
Smithsonian Institution 210–211
Smith Woodward, Arthur 150, 162
Smuts, Jan Christian 163–164
Snake Park, Port Elizabeth Museum
    46, 47
social Darwinism 156, 166–167
Society of Africanist Archaeologists
    242–243
Soudien, Crain 284–285
South African Association for the
    Advancement of Science (S2A3)
    14–18, 103, 110–111, 114, 135,
    151–155, 163–164, 167, 174
South African College, Medical School
    21, 25, 29, 65, 76–77
South African Cultural History Museum
    271
'South Africanisation' of science 135–136,
    268
South African Journal of Science 292
South African Medical and Dental
    Council (SAMDC) 239–240
South African Medical Corps 75
South African Medical Journal (SAMJ)
    5–6, 286–289
South African Medical Record 82
South African Museum *19*, 35, 84, 123,
    148, 217–218, 270–271

South African Philosophical Society 15
South African War 14–16, 112, 124, 248
South West Africa 18, 156, 173–175, 217
Spencer, Frank 191–192
    *Encyclopaedia of the History of
        Physical Anthropology* 191–192
Sperber, Geoff 229
splitters 273
Springbok Flats 54–55, 58–59
Star, The 52, 58
statistical analyses 276–277
Statistics South Africa 286
'Stellenbosch' 12
Stepan, Nancy 221, 294
Sterkfontein fossil site *198*
Stevens, Royal 157
Stewart, T. Dale 206, 230
Stibbe, Edward 146, 147, 149
stone cairns 121, 133
Strandlopers xix, 23, 30–31, 50–51, 78, 83
structural-functionalist school of social
        anthropology 166
Stuttgart, SS 140
Summers, Roger 11
Supreme Court 252–253, 261–263, 265
surgeons 66
Suzman, Helen 182, 240
Suzman, Ivan *240*
Swiss Institute of Anthropology 34

T
Taung skull 58, 150–155, *151*, 160,
        162–163, 181, 183, 186
Teitz, Nancy 45
Terry, Robert J. 145
Thomson, Robert Black 25, 43–44, 48, 65,
        72, 74–75, 145
Thorne, Charles 36
Tobias, Phillip
    anthropology, interest in 205
    background of xvii, 193–197
    *Bantu-Speaking Peoples of Southern
        Africa, The* 224–225
    biometrical analysis 204–205
    Boskop type 56
    Campbell cemetery *238*
    Dart and 149, 180, 224, *240*, 269
    De Villiers and 232, 237–239
    Drennan and 74
    Galloway and 160, 178
    'History of Physical Anthropology
        in Southern Africa' (paper) 3,
        192–193
    Makapansgat site 197–199
    management style 226–229, 238–239
    Meaning of Race, The (pamphlet) 224
    'new physical anthropology' 57, 186,
        191–192, 223, 243, 273
    Nuffield Senior Fellowship 203–205
    Panhard- Capricorn Expedition
        199–203, *201*
    Into the Past 193
    political views 9, 194–196, 219, 224,
        226, 236, 239–243, 247
    publications of 193
    race classification 264–265, 267
    research interests 232–237
    Rockefeller Travelling Fellowship 203,
        205–207
    Singer and 219
    supervisions 229–230
    Weiner and 223
    Wells and 173
    at Wits 172, 199–203, 225–230,
        232–243
Tobias, Valerie 194
Todd, T. Wingate 165
tombs 61, *62*
training *see* medical education
*Transactions of the Royal Society of South
        Africa* 44, 78
Transvaal Museum 41–43, 58–59
Trevor, J.C. (Jack) 203–204, 230
Trevor-Jones, Trevor *169*, 265
Tricameral Parliament 251
Trimen, Director 12
Tsitsikamma region 49–55, 61, *62*,
        147–149
Tucker, Benjamin 240
Tucker, Robert *19*
Tucker, Winnifred 77
Turner, William 70–72

typology
  apartheid and 260, 268–269
  Broom, Robert 130–132
  current views 290–291, 294–295
  Dart, Raymond 167, 170–186,
    224–225
  Drennan, Matthew Robertson 86
  history of 4–6, 109–112, 121, 221, 284
  Hoffman, Abraham 125–126
  Mendelian geneticists 220
  nationalism and 132–142
  racist implications of 179–186
  Tobias, Phillip 192, 203, 224

U
Ucko, Peter 241–242
UCT *see* University of Cape Town
Udusa *see* Union of Democratic
    University Staff Associations
Uganda 178, 232–233
Unesco *Statement on Race* 202, 204,
    220–221, 224, 272, 279
Union of Democratic University Staff
    Associations (Udusa) 242
Union of South Africa 15, 248
United Empire Exhibition (1937) 89–90
United Nations Educational, Scientific
    and Cultural Organization *see*
    Unesco *Statement on Race*
United States *see* America
University Act 74
University College (later University of the
    Witwatersrand) 29
University College, London 144
University of Birmingham 163–164
University of California African
    Expedition 175
University of Cape Town (UCT) 4–8, 21,
    74–75, 79, 86, 103, 148, 167, 208,
    271–273
University of Cernauti 160
University of Halle-Wittenberg (later
    Martin Luther University)
    112–113
University of Stellenbosch 137–138, 140,
    213–214, 291–292

University of the Cape of Good Hope 12,
    65, 93
University of the Witwatersrand (Wits)
    47, 148, 159–160, 207, 225–243
Upper Palaeolithic 12–14
urine sampling 234

V
Van Hoepen, Egbert Cornelis Nicolaas
    xvi, 124–126, 134–135, 269
Van Niekerk, Anton 287–288
Van Riet Lowe, Clarence 14, 54, 119, 134,
    172, 197, 199
Van Zyl, David 270–271
Vedder, Heinrich 133
  South West Africa in Early Times 133
Vermooten, Vincent 77–79, 83
Verwoerd, Hendrik 140
vineyards 11–12
volkekunde 140, 268, 269
Von Bonde, W. 116
Von Luschan, Felix xvi, 16–19, 22,
    110–111, 137
Von Schlaginhaufen, Otto 137
Vorster, Willie 65–66

W
Walters, Handri 138–139
Washburn, Sherwood (Sherry) 206,
    221–222
Weiner, Joseph (Joe) xvii, 149, 207, 212,
    217, 222–224, 226, 234, 279
  *Peoples of Southern Africa and Their
    Affinities, The* 279
Weingart, Peter 293–294
Wells, Lawrence
  background of xvii, 172–173
  Boskop type 54
  craniology 121, 178, 230
  Dart and 149, 171–172, 177–178, 186
  eugenics movement 140
  Kakamas type 133
  Mapungubwe site 168, *169*
  personality of 175
  political views 270
  publications of 176–178

Wells, Lawrence (*continued*)
    race classification 265, 267
    research interests 176–178
    typology 179
    at UCT 176, 178–179
    at University of Edinburgh 176–177,
      199
    at Wits 172–176, 235
Wessels, Quenton 65–66
Wheelhouse, Frances 157, 180
Whitcher's Cave *49*, 172
Wikar, Hendrik Jacob 126–127
Wilder, Harris 137
    Laboratory Manual of Anthropometry
      137
Williams, Eric W. 156, 175, 202
Wilson, James 144, 145, 163
'Wilton Culture' 117

Wits *see* University of the Witwatersrand
Witwatersrand Central School Board 263
Woods Hole Research Institute,
    Massachusetts 145
Word War II *see* Second World War
World Archaeological Bulletin 280
World Archaeology Congress 241–243
Wymer, John 218

Z
Zeitschrift für Rassenkunde 136–141
Zieve, Isidore 89
Zimbabwe ruins *see* Great Zimbabwe
    ruins
Zoological Society of London 163
Zuckerman, Solly xvii, 84–85, 204
Zululand expedition 155–156
Zuurberg site 172

Lightning Source UK Ltd.
Milton Keynes UK
UKHW012110220322
400452UK00001B/51